HAIL ALLEY

WEATHER WARRIORS

by Alexander Lowe • illustrated by Sebastian Kadlecik

NORWOOD
DISCOVERY
Graphics

Norwood House Press

For more information about Norwood House Press please visit our website at: www.norwoodhousepress.com or call 866-565-2900.

Library of Congress Cataloging-in-Publication Data
Names: Lowe, Alexander, author. | Kadlecik, Sebastian, illustrator.
Title: Hail alley / by Alexander Lowe ; illustrated by Sebastian Kadlecik.
Description: Chicago : Norwood House Press, [2021] | Series: Norwood discovery graphics | Audience: Ages 8-10 |
    Audience: Grades 4-6 | Summary: "Follow a family as they vacation in Colorado during a hail storm. An adventure-filled
    graphic novel that provides young readers information about hail and related weather conditions. Learn from these
    weather warriors the early signs of a hail storm, as well as how to protect yourself and property during a storm. Includes
    contemporary full-color graphic artwork, fun facts, additional information, and a glossary"— Provided by publisher.
Identifiers: LCCN 2020024512 (print) | LCCN 2020024513 (ebook) | ISBN 9781684508549
    (hardcover) | ISBN 9781684045938 (paperback) | ISBN 9781684045983 (epub)
Subjects: LCSH: Hailstorms—Comic books, strips, etc. | Hailstorms—Juvenile literature. | Hailstorms—
    Colorado—Juvenile literature. | Graphic novels. | CYAC: Graphic novels.
Classification: LCC QC929.H15 L69 2021 (print) | LCC QC929.H15 (ebook) | DDC 551.55/4—dc23
LC record available at https://lccn.loc.gov/2020024512
LC ebook record available at https://lccn.loc.gov/2020024513

Hardcover ISBN: 978-1-68450-854-9   Paperback ISBN: 978-1-68404-593-8

328N—072020
Manufactured in the United States of America in North Mankato, Minnesota.

CONTENTS

MEET THE WEATHER WARRIORS

Kim

Steve

Frankie

Sarah

Olivia

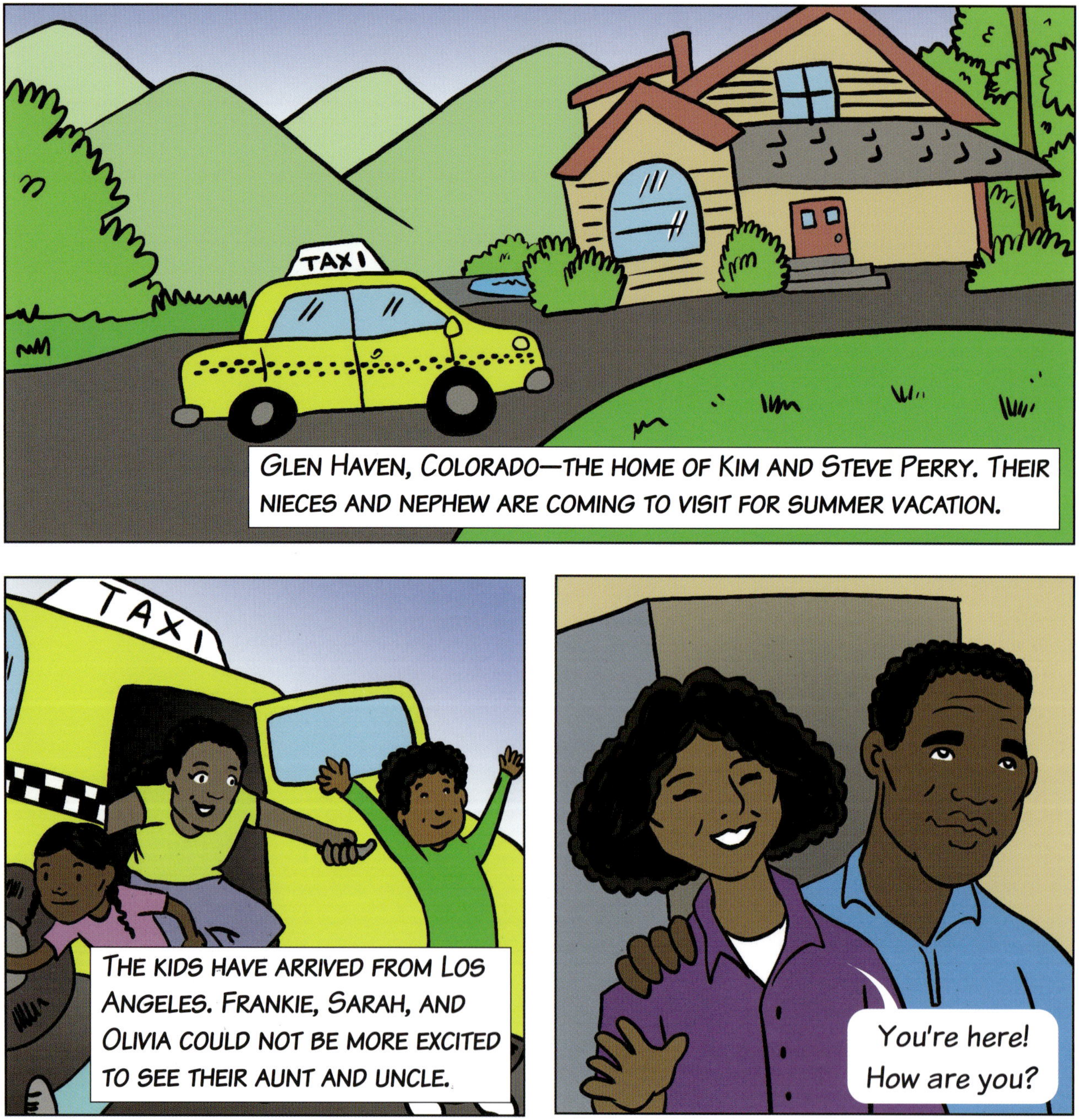
TAXI
GLEN HAVEN, COLORADO—THE HOME OF KIM AND STEVE PERRY. THEIR NIECES AND NEPHEW ARE COMING TO VISIT FOR SUMMER VACATION.
TAXI
THE KIDS HAVE ARRIVED FROM LOS ANGELES. FRANKIE, SARAH, AND OLIVIA COULD NOT BE MORE EXCITED TO SEE THEIR AUNT AND UNCLE.
You're here! How are you?

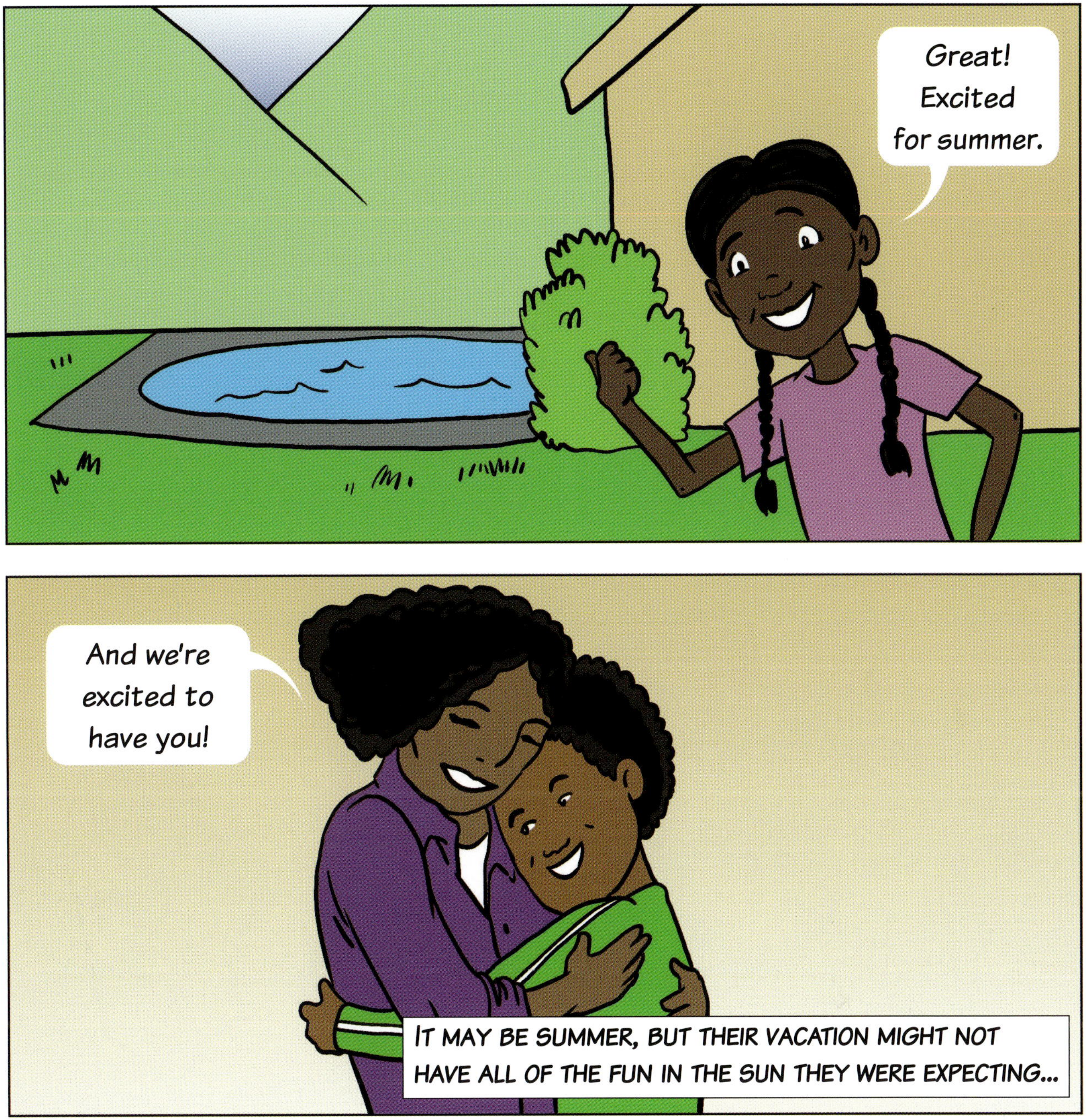
Great! Excited for summer.
And we're excited to have you!
IT MAY BE SUMMER, BUT THEIR VACATION MIGHT NOT HAVE ALL OF THE FUN IN THE SUN THEY WERE EXPECTING...

A HAILSTORM IS ON THE WAY.

What do you guys want to do today?

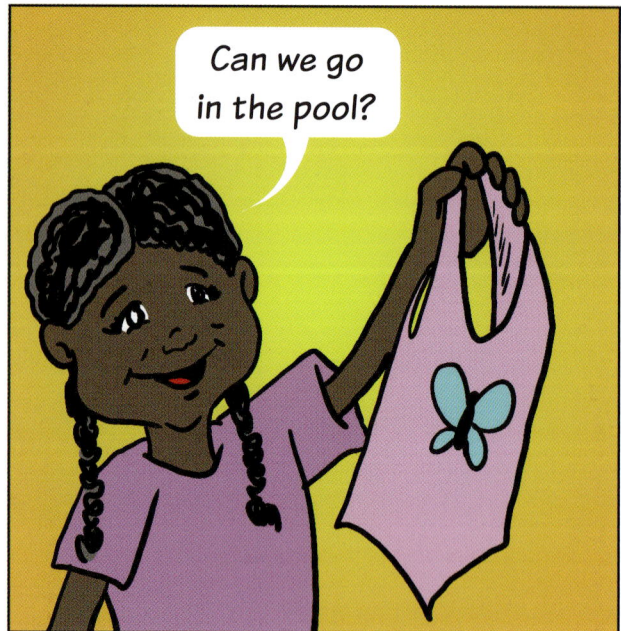

Can we go in the pool?

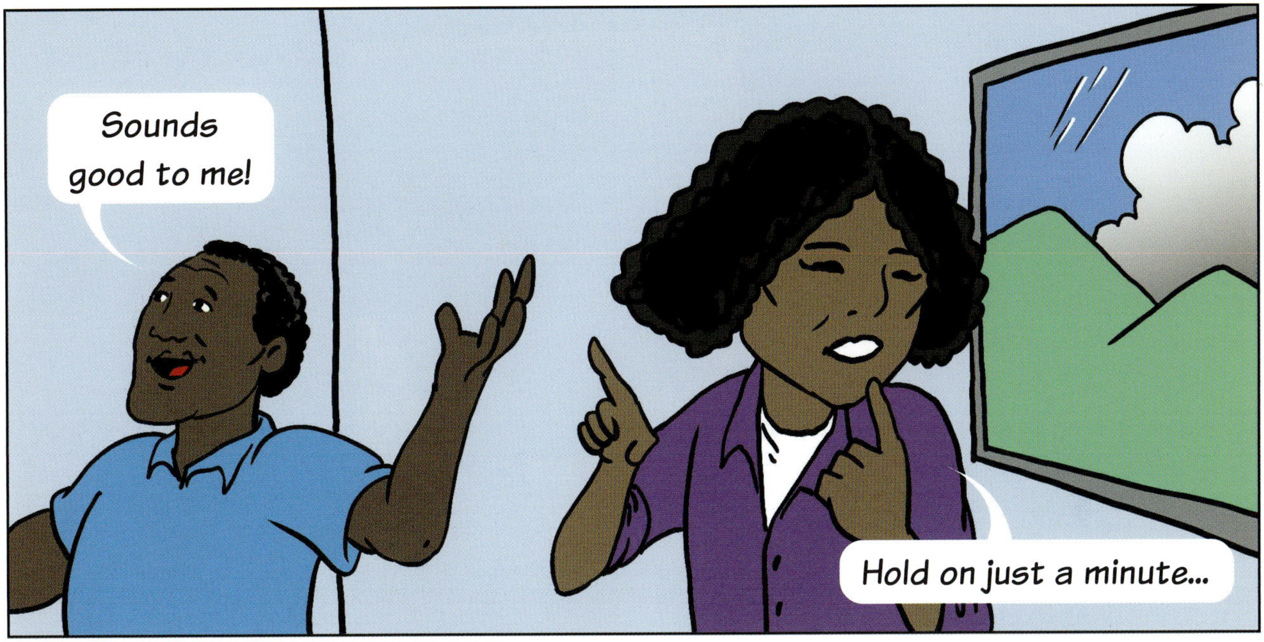
Sounds good to me!
Hold on just a minute...

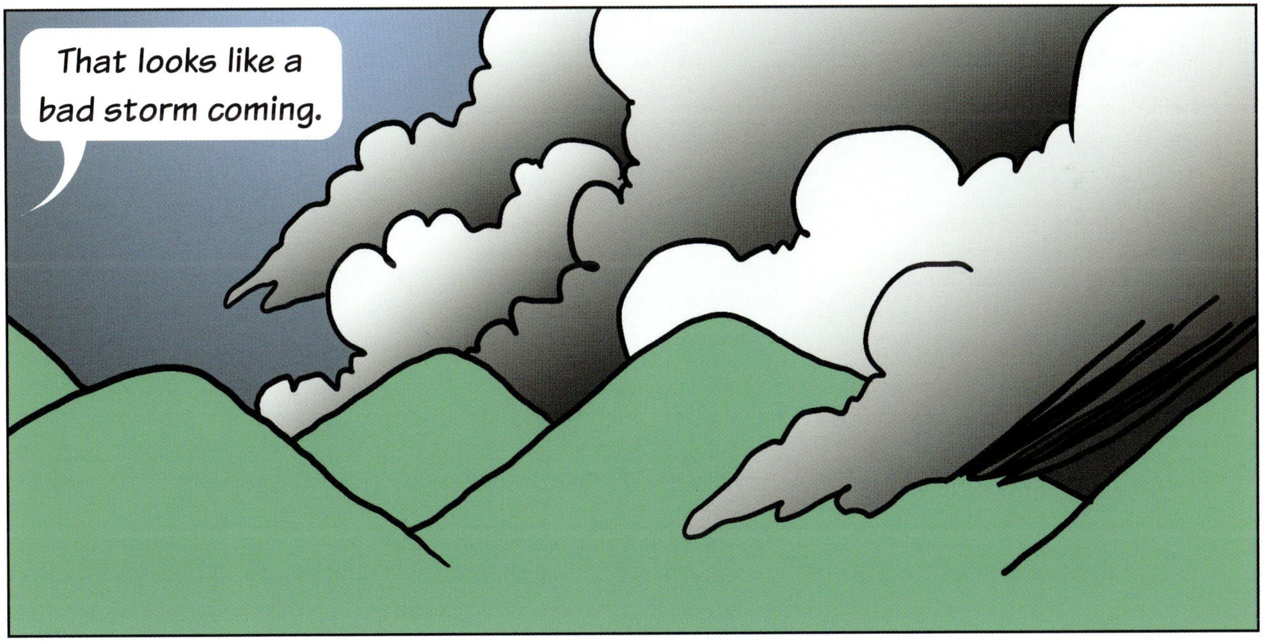
That looks like a bad storm coming.

All Kinds of Precipitation

Hail is a type of **precipitation**, but it's not the only one. All precipitation comes from water vapor. The most common are rain and snow. There is also sleet. Sleet looks like smaller, wetter hail. Sleet forms when snow falls into a layer of air that is warm. The snow melts into rain. Then that rain falls into a layer of air that is cold. Then it refreezes and makes sleet.

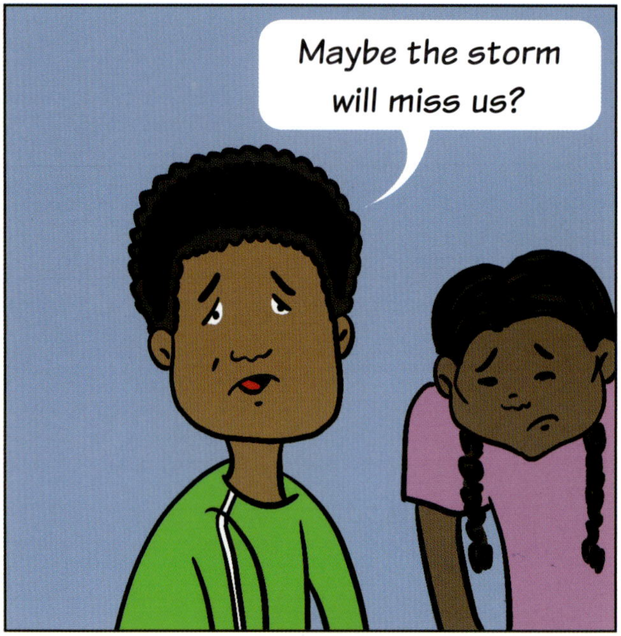
Maybe the storm will miss us?

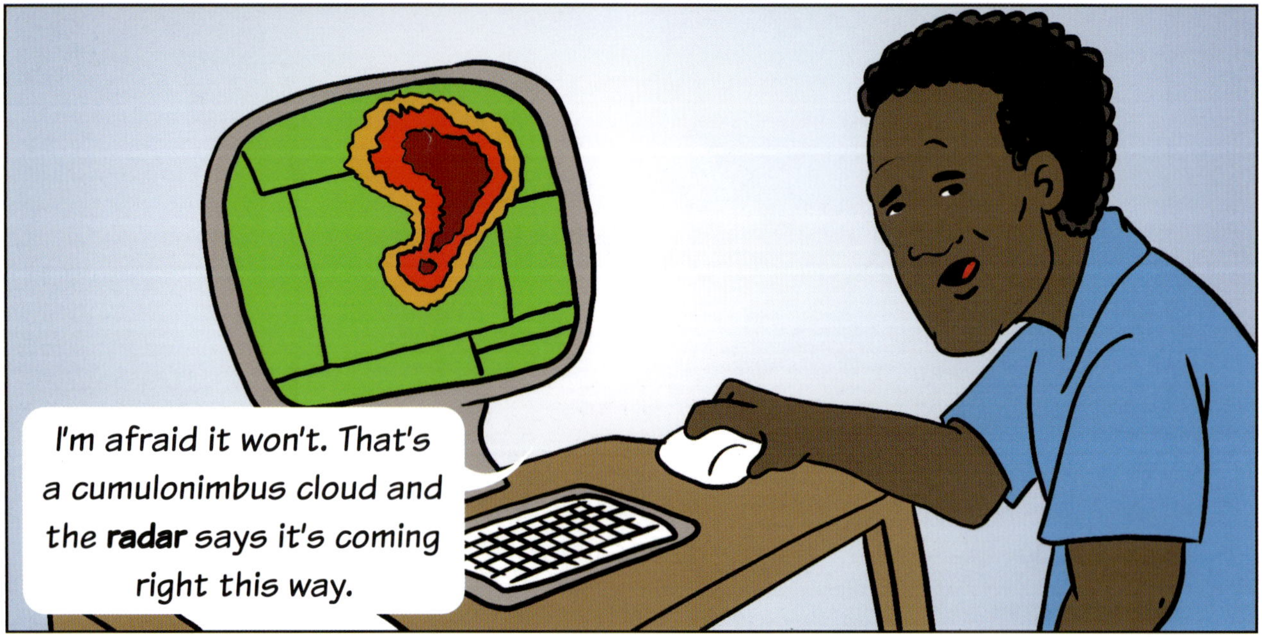
I'm afraid it won't. That's a cumulonimbus cloud and the radar says it's coming right this way.

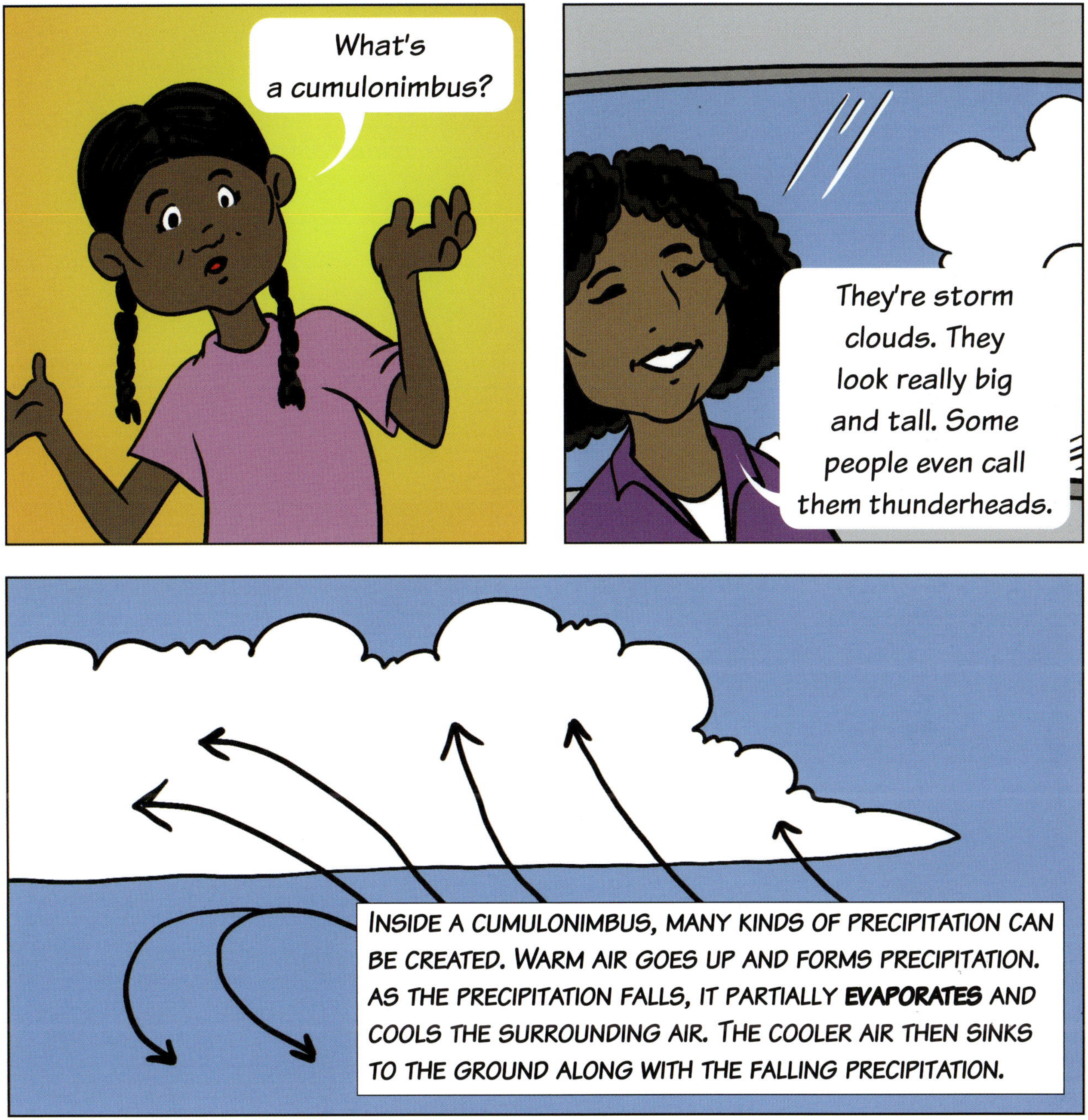
What's a cumulonimbus?
They're storm clouds. They look really big and tall. Some people even call them thunderheads.
INSIDE A CUMULONIMBUS, MANY KINDS OF PRECIPITATION CAN BE CREATED. WARM AIR GOES UP AND FORMS PRECIPITATION. AS THE PRECIPITATION FALLS, IT PARTIALLY **EVAPORATES** AND COOLS THE SURROUNDING AIR. THE COOLER AIR THEN SINKS TO THE GROUND ALONG WITH THE FALLING PRECIPITATION.

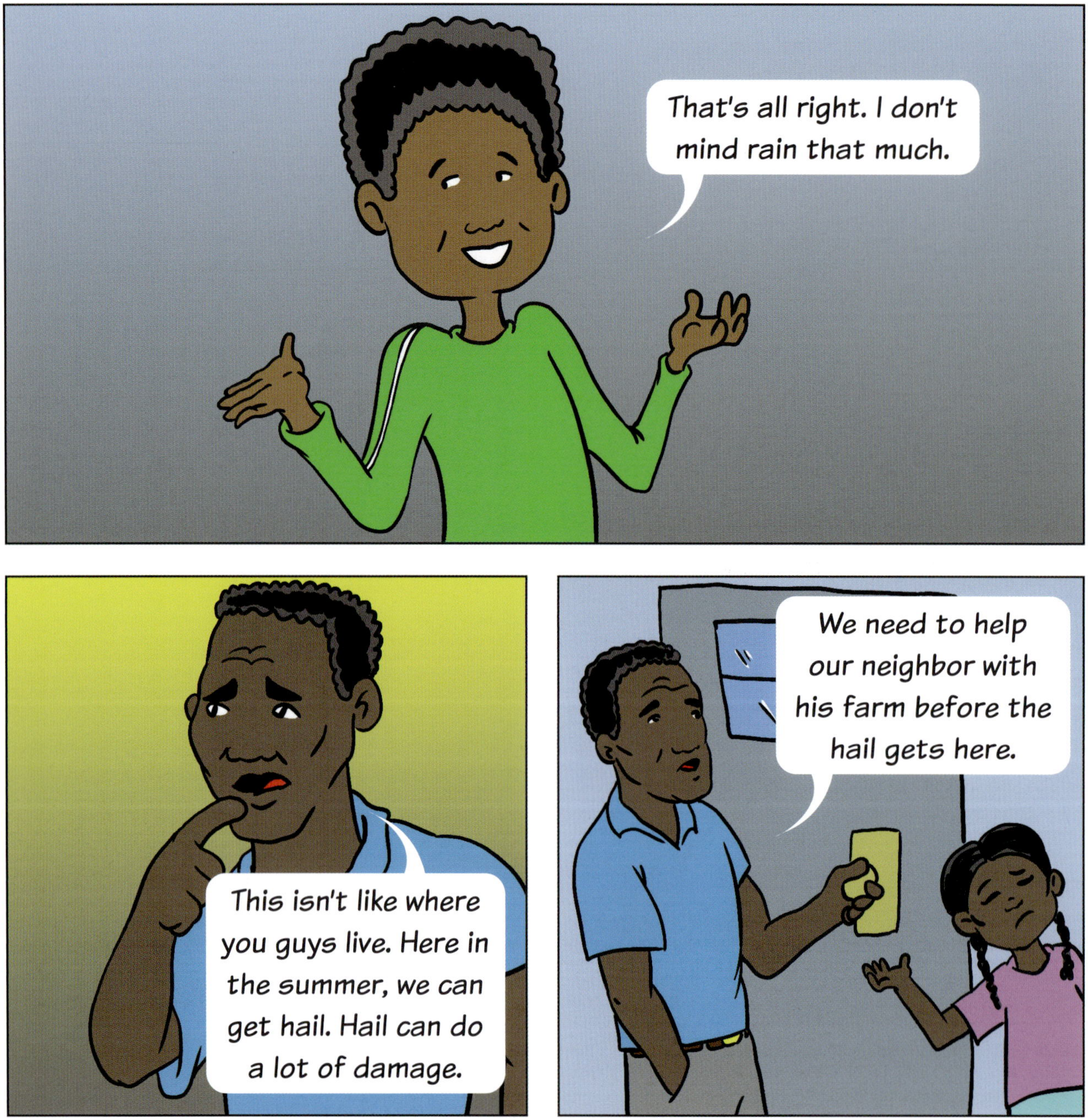
That's all right. I don't mind rain that much.
This isn't like where you guys live. Here in the summer, we can get hail. Hail can do a lot of damage.
We need to help our neighbor with his farm before the hail gets here.

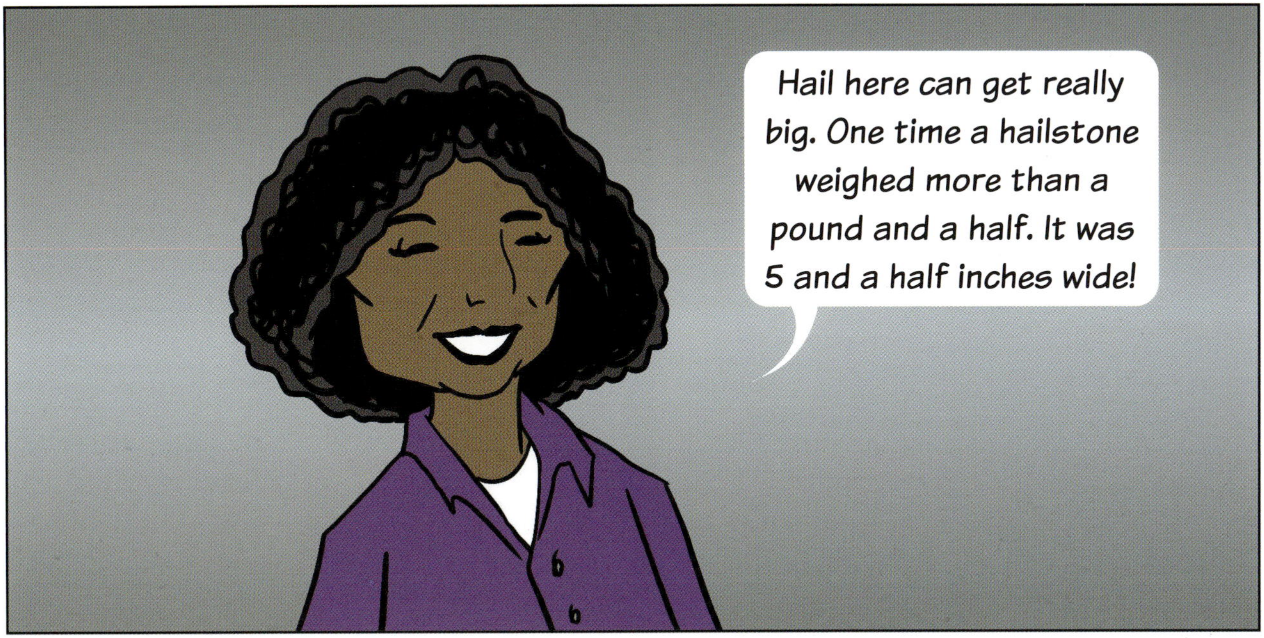
Hail here can get really big. One time a hailstone weighed more than a pound and a half. It was 5 and a half inches wide!

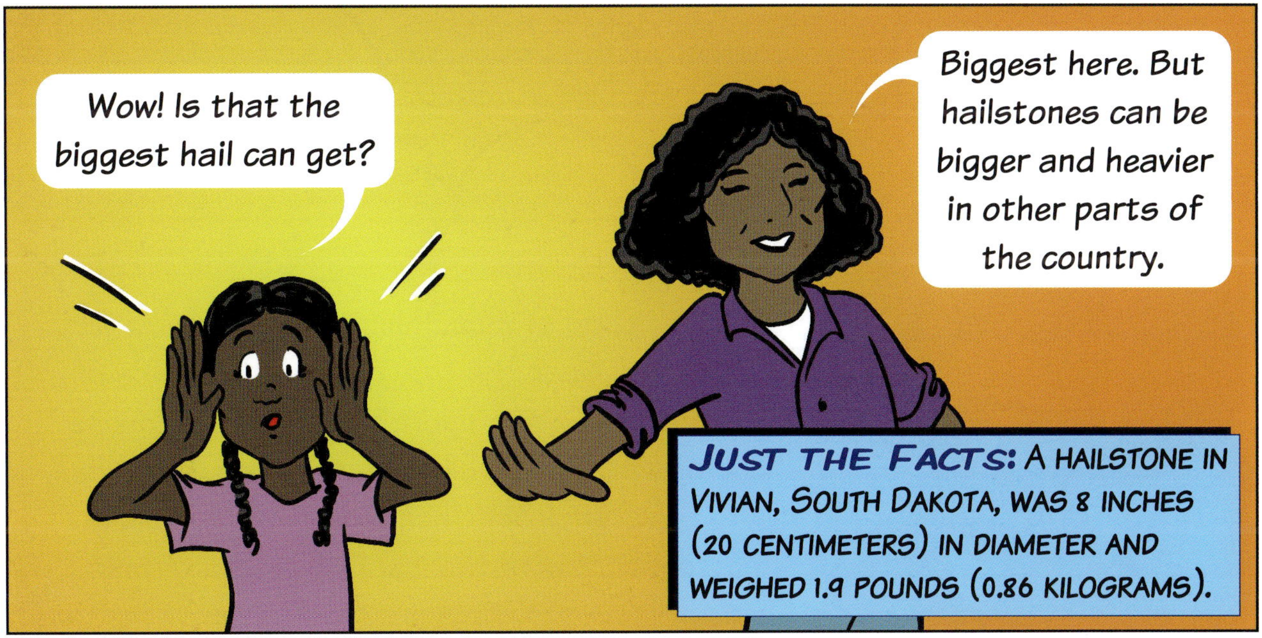
Wow! Is that the biggest hail can get?
Biggest here. But hailstones can be bigger and heavier in other parts of the country.
JUST THE FACTS: A HAILSTONE IN VIVIAN, SOUTH DAKOTA, WAS 8 INCHES (20 CENTIMETERS) IN DIAMETER AND WEIGHED 1.9 POUNDS (0.86 KILOGRAMS).

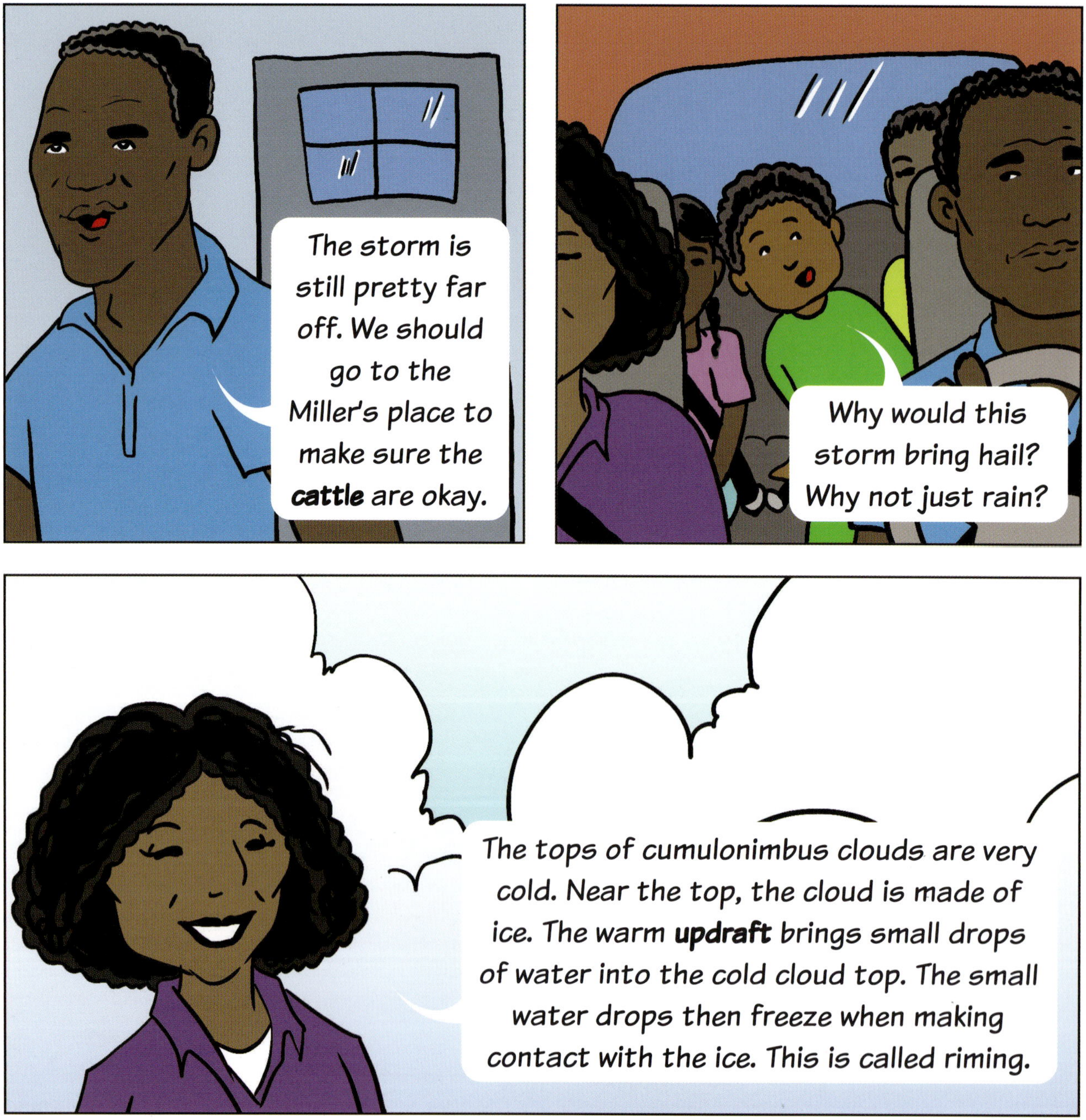
The storm is still pretty far off. We should go to the Miller's place to make sure the **cattle** are okay.
Why would this storm bring hail? Why not just rain?
The tops of cumulonimbus clouds are very cold. Near the top, the cloud is made of ice. The warm **updraft** brings small drops of water into the cold cloud top. The small water drops then freeze when making contact with the ice. This is called riming.

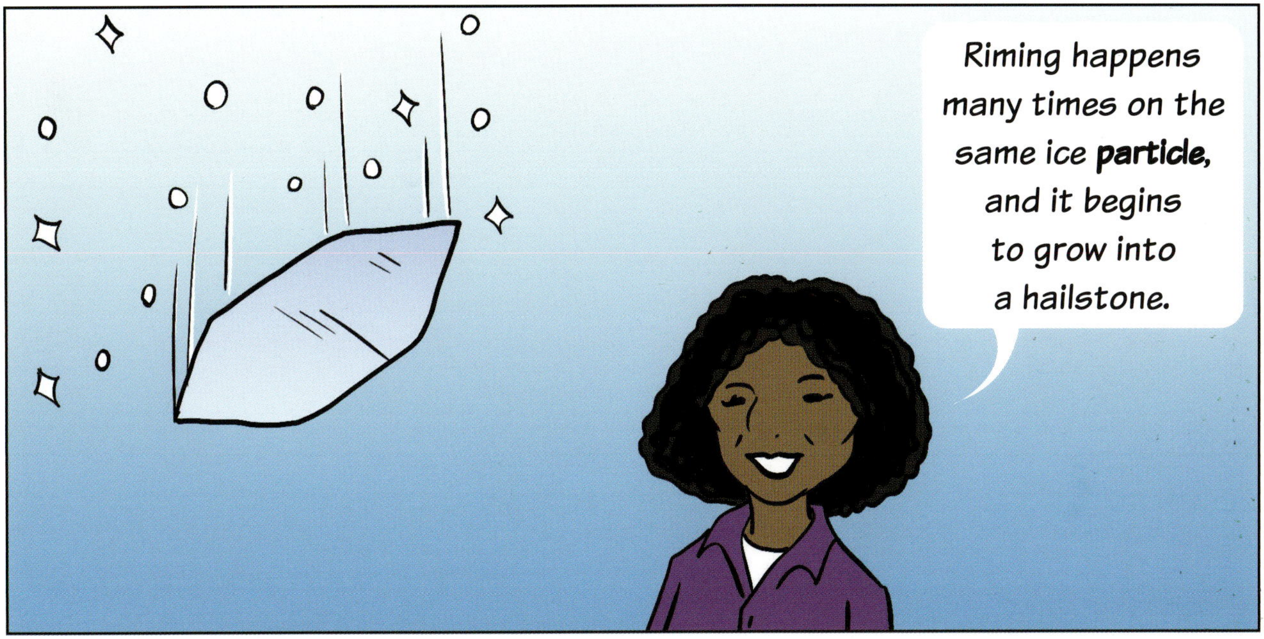
Riming happens many times on the same ice **particle**, and it begins to grow into a hailstone.

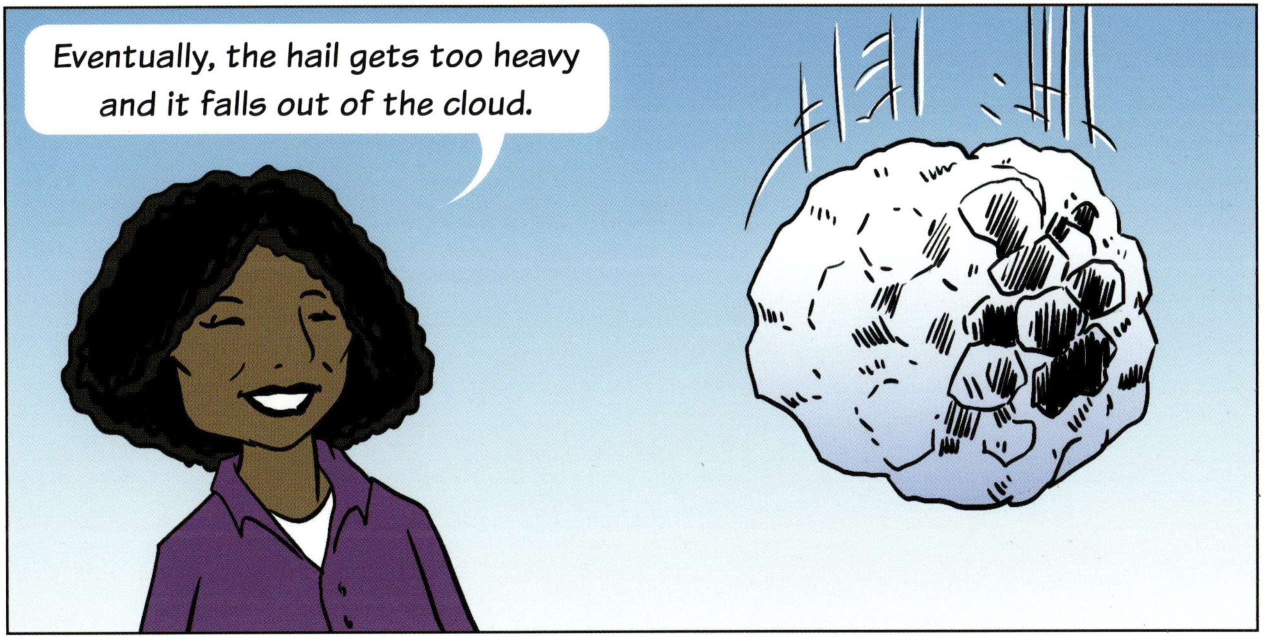
Eventually, the hail gets too heavy and it falls out of the cloud.

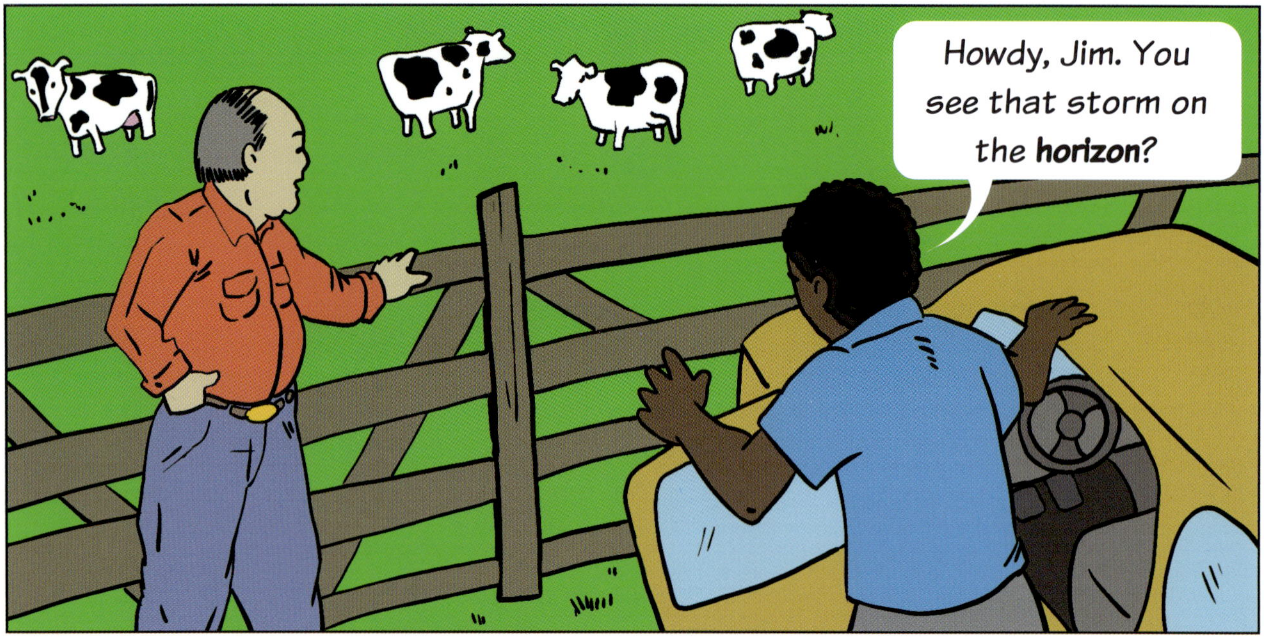
Howdy, Jim. You see that storm on the **horizon**?

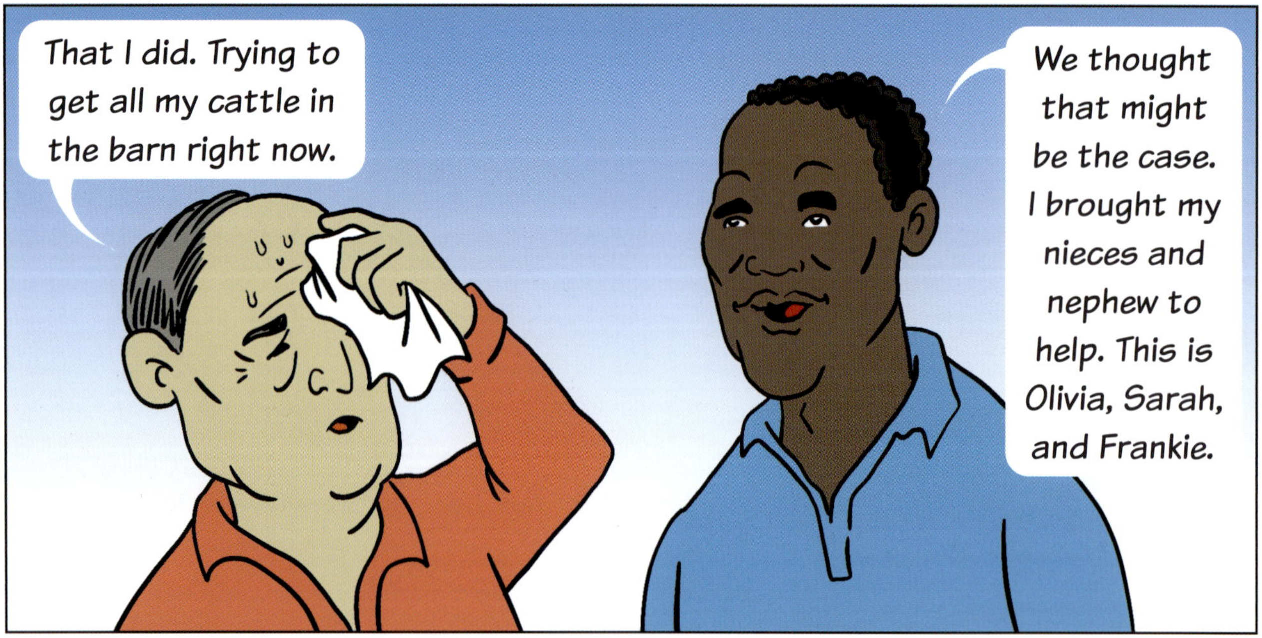
That I did. Trying to get all my cattle in the barn right now.
We thought that might be the case. I brought my nieces and nephew to help. This is Olivia, Sarah, and Frankie.

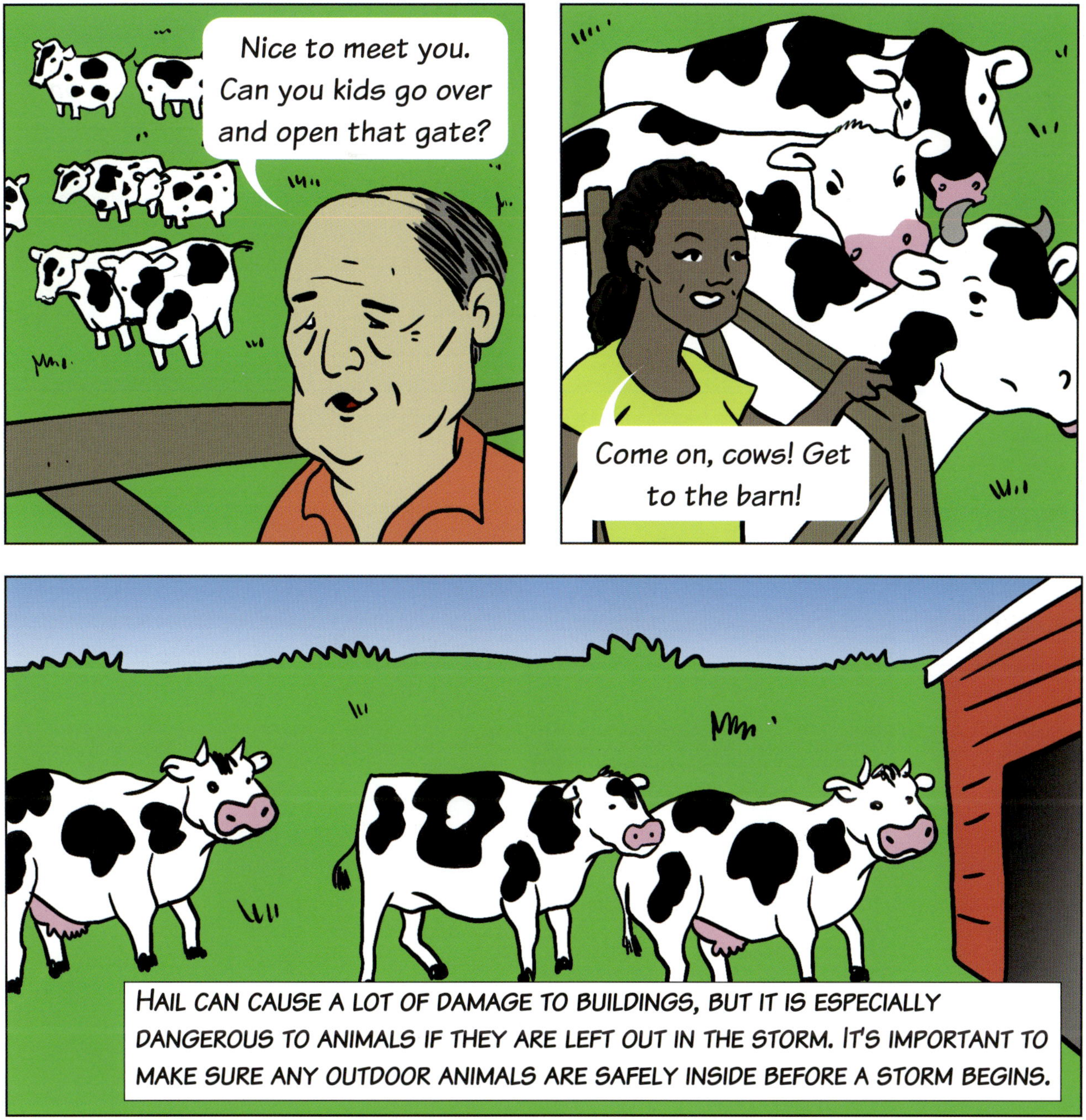
Nice to meet you. Can you kids go over and open that gate?
Come on, cows! Get to the barn!
HAIL CAN CAUSE A LOT OF DAMAGE TO BUILDINGS, BUT IT IS ESPECIALLY DANGEROUS TO ANIMALS IF THEY ARE LEFT OUT IN THE STORM. IT'S IMPORTANT TO MAKE SURE ANY OUTDOOR ANIMALS ARE SAFELY INSIDE BEFORE A STORM BEGINS.

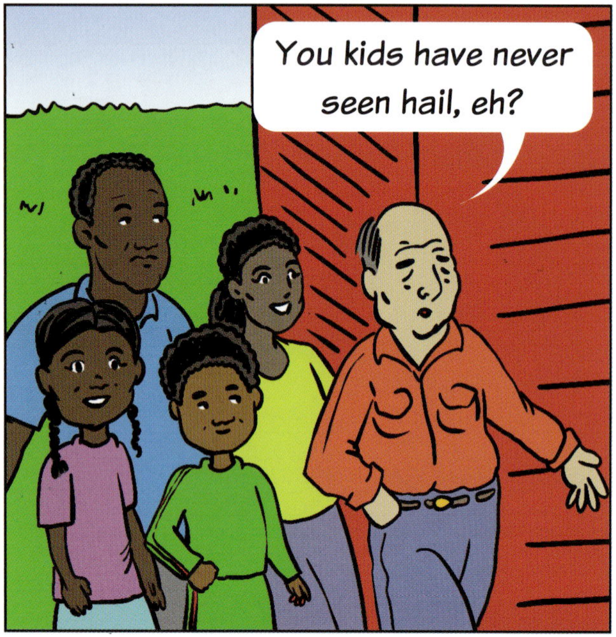
You kids have never seen hail, eh?

HUGE HAILSTORMS

There have been many large hailstorms. Some of them have caused a lot of damage. This can lead to a high cost for repairs. In 2017, Colorado got hit hard by a hailstorm. The total damage was estimated to be $2.3 billion. In 2012, St. Louis, Missouri, was hit with a series of hailstorms. Those led to $1.6 billion in damage.

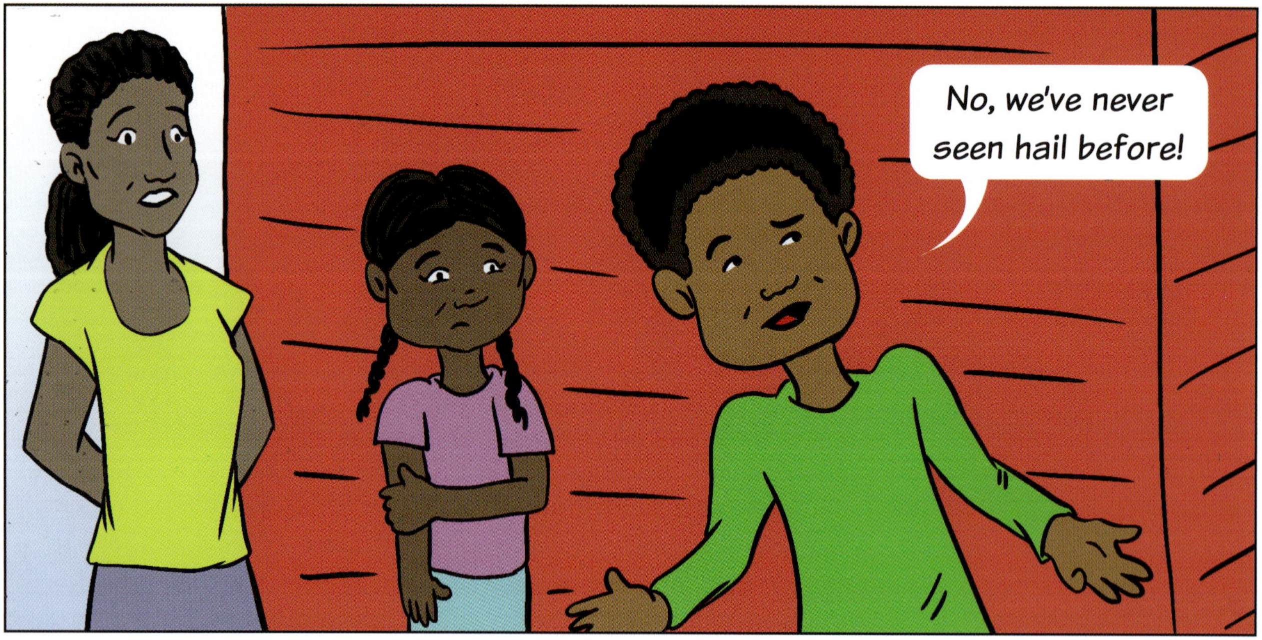
No, we've never seen hail before!

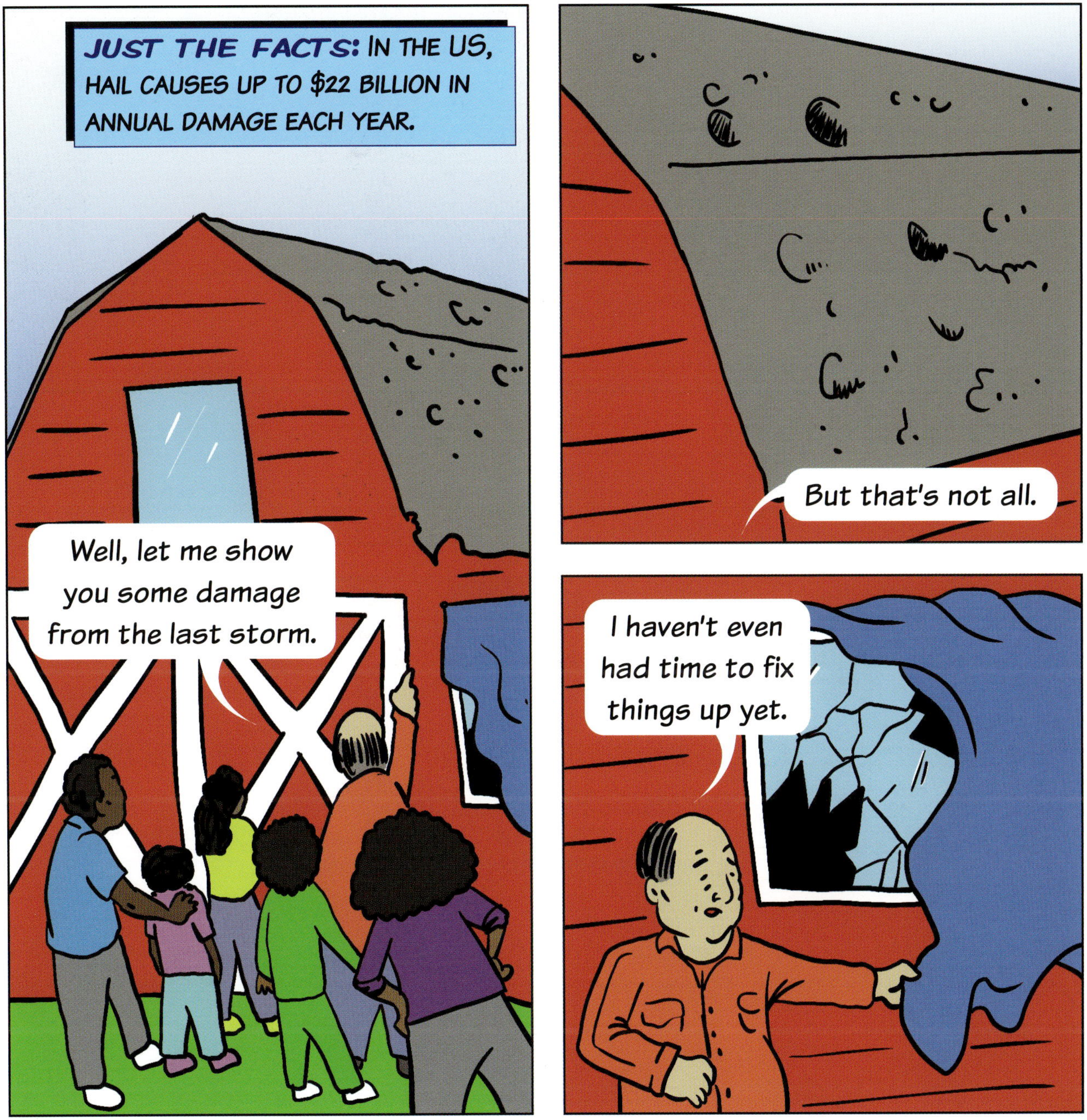
JUST THE FACTS: IN THE US, HAIL CAUSES UP TO $22 BILLION IN ANNUAL DAMAGE EACH YEAR.
Well, let me show you some damage from the last storm.
But that's not all.
I haven't even had time to fix things up yet.

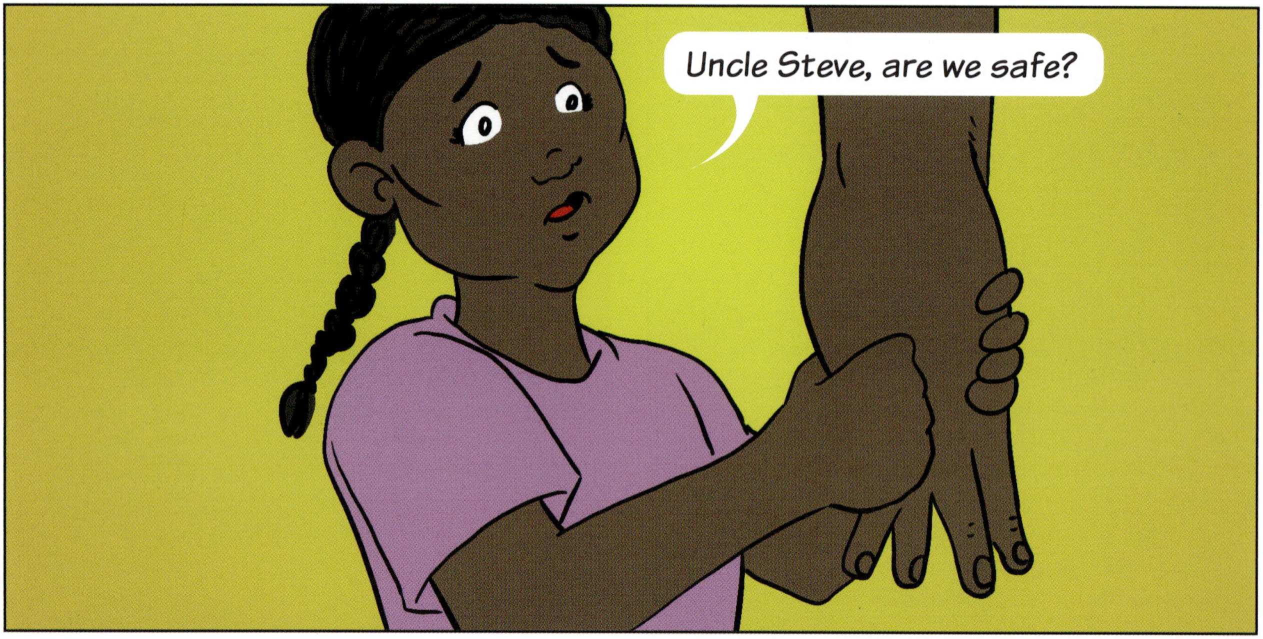
Uncle Steve, are we safe?

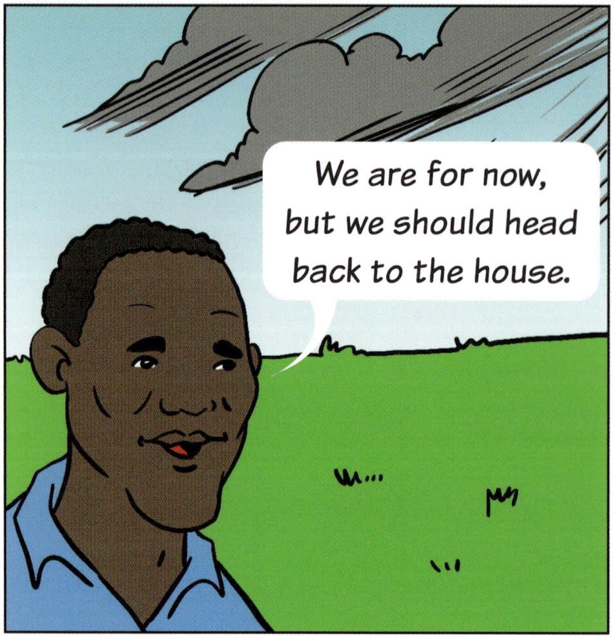
We are for now, but we should head back to the house.

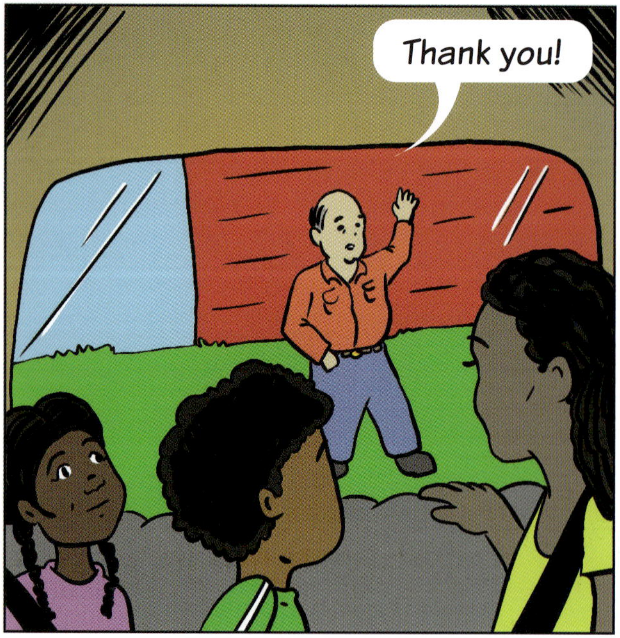
Thank you!

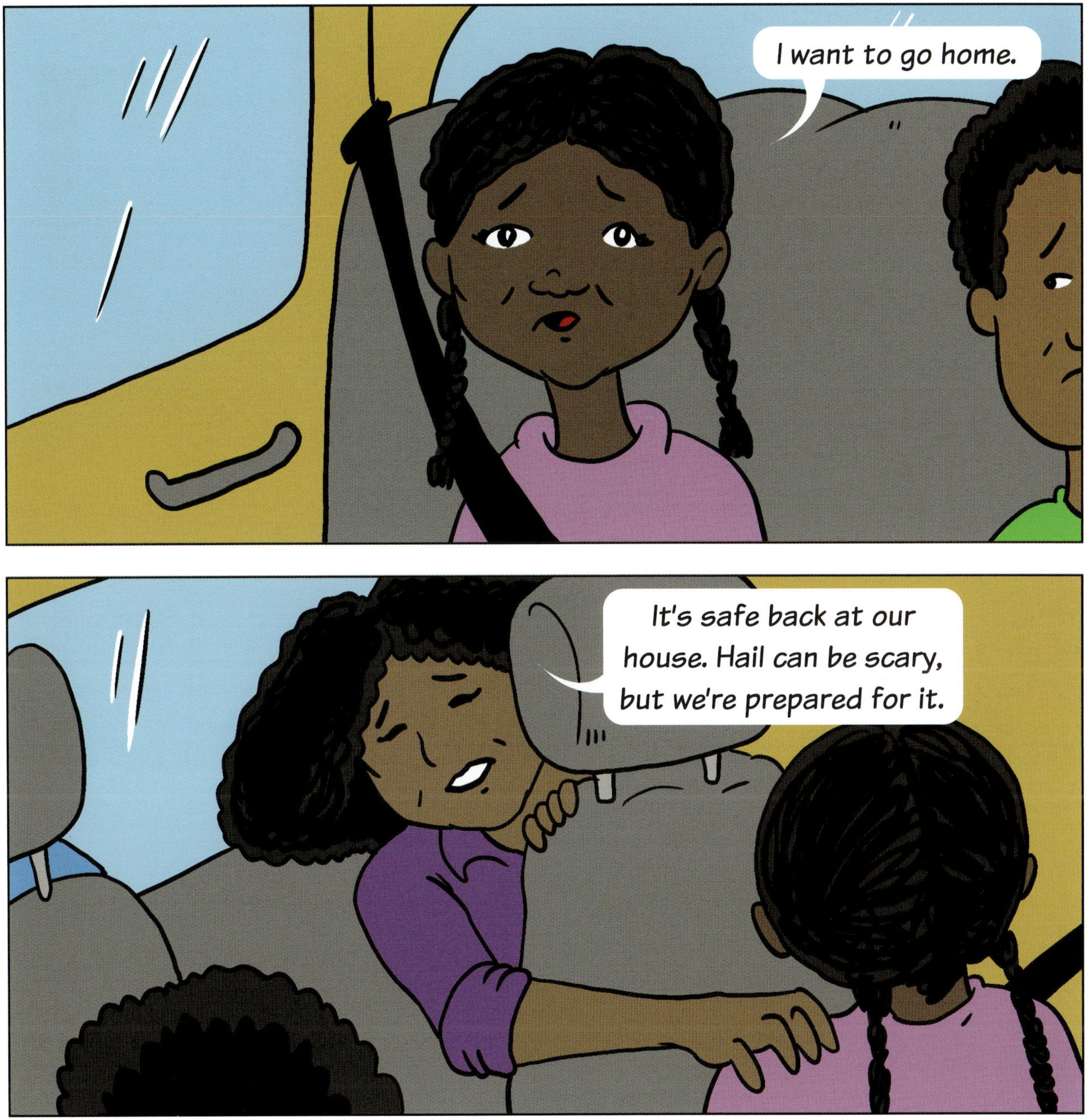
I want to go home.
It's safe back at our house. Hail can be scary, but we're prepared for it.

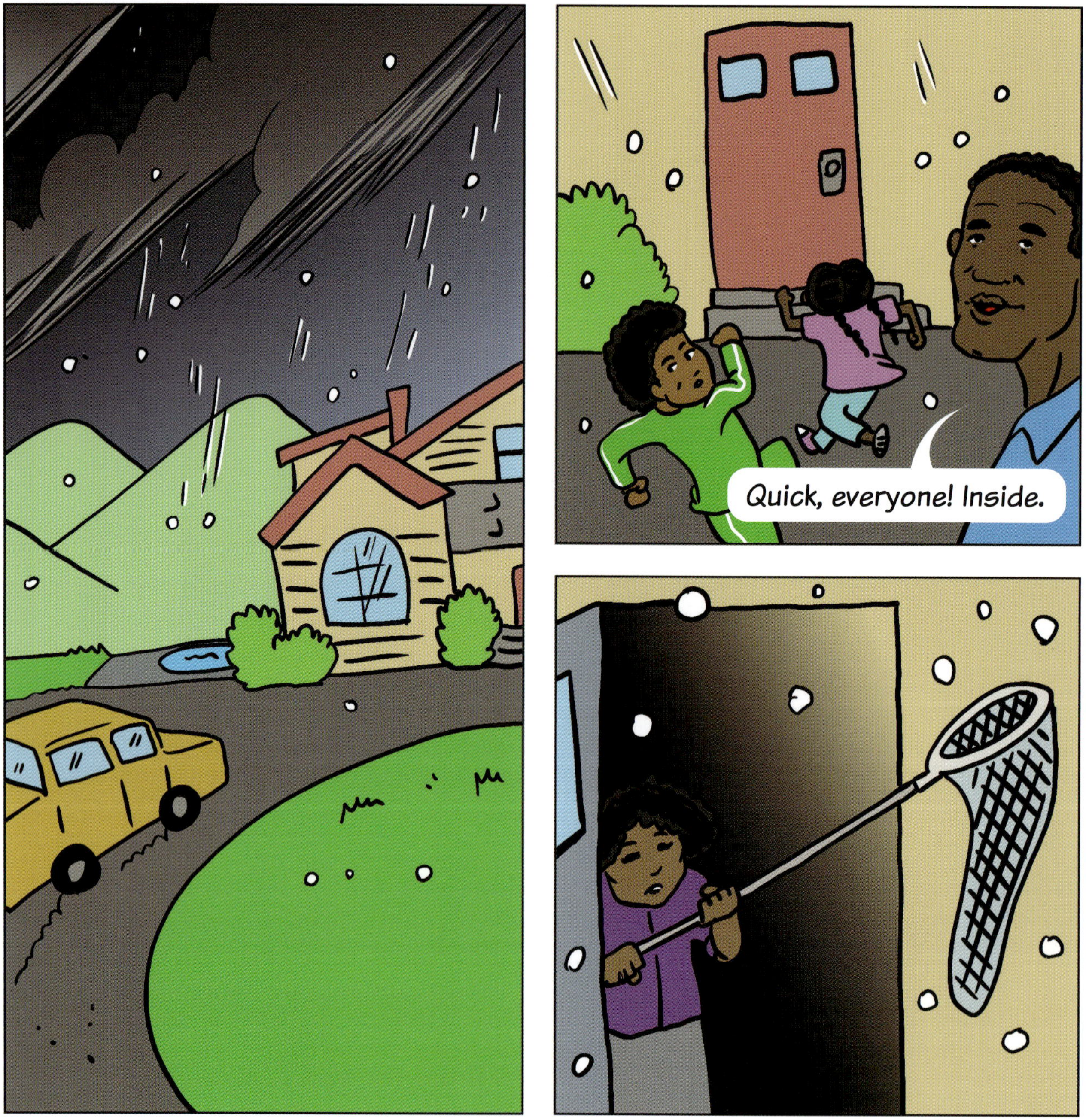
Quick, everyone! Inside.

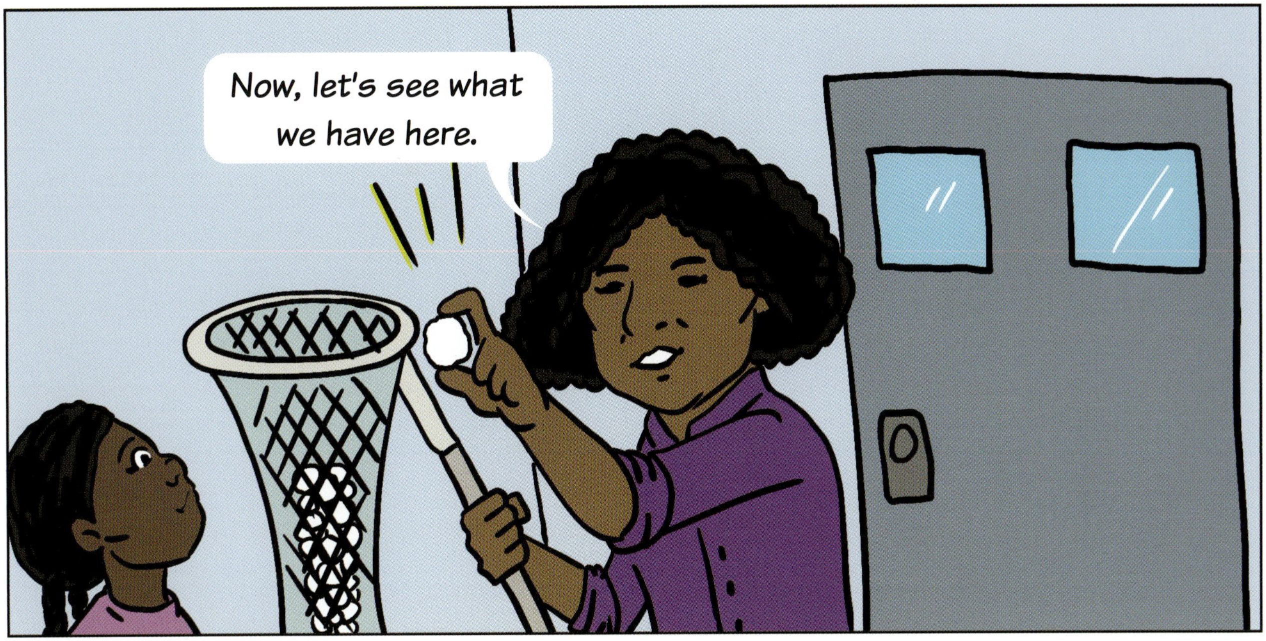
Now, let's see what we have here.

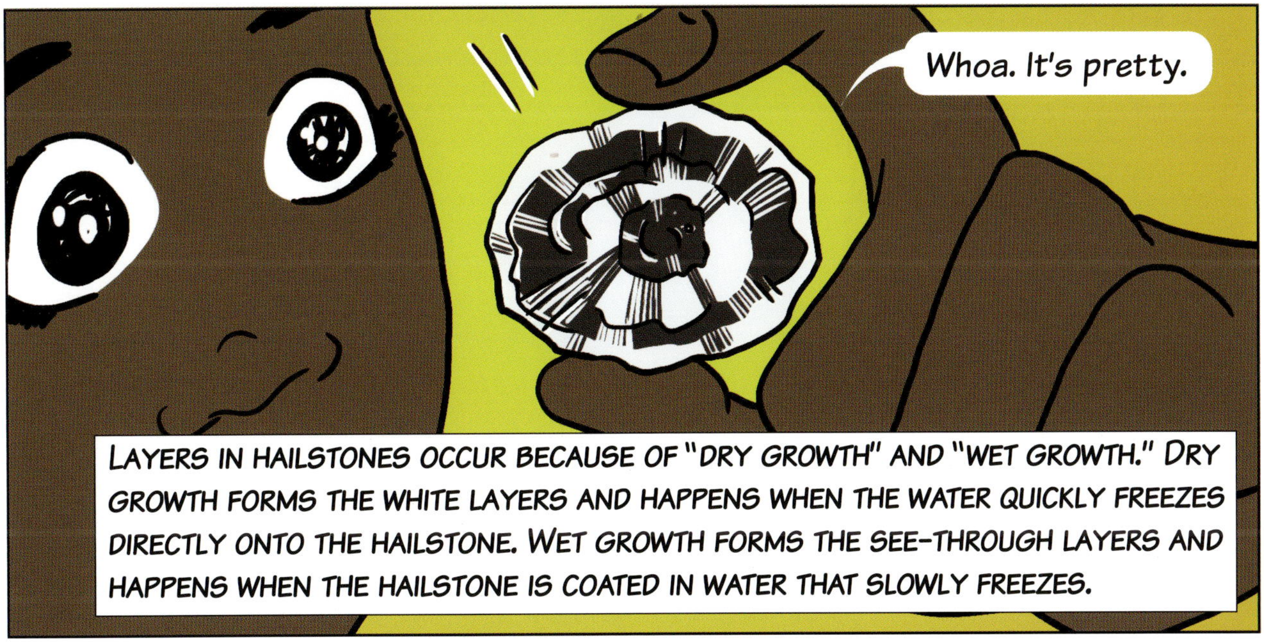
Whoa. It's pretty.
LAYERS IN HAILSTONES OCCUR BECAUSE OF "DRY GROWTH" AND "WET GROWTH." DRY GROWTH FORMS THE WHITE LAYERS AND HAPPENS WHEN THE WATER QUICKLY FREEZES DIRECTLY ONTO THE HAILSTONE. WET GROWTH FORMS THE SEE-THROUGH LAYERS AND HAPPENS WHEN THE HAILSTONE IS COATED IN WATER THAT SLOWLY FREEZES.

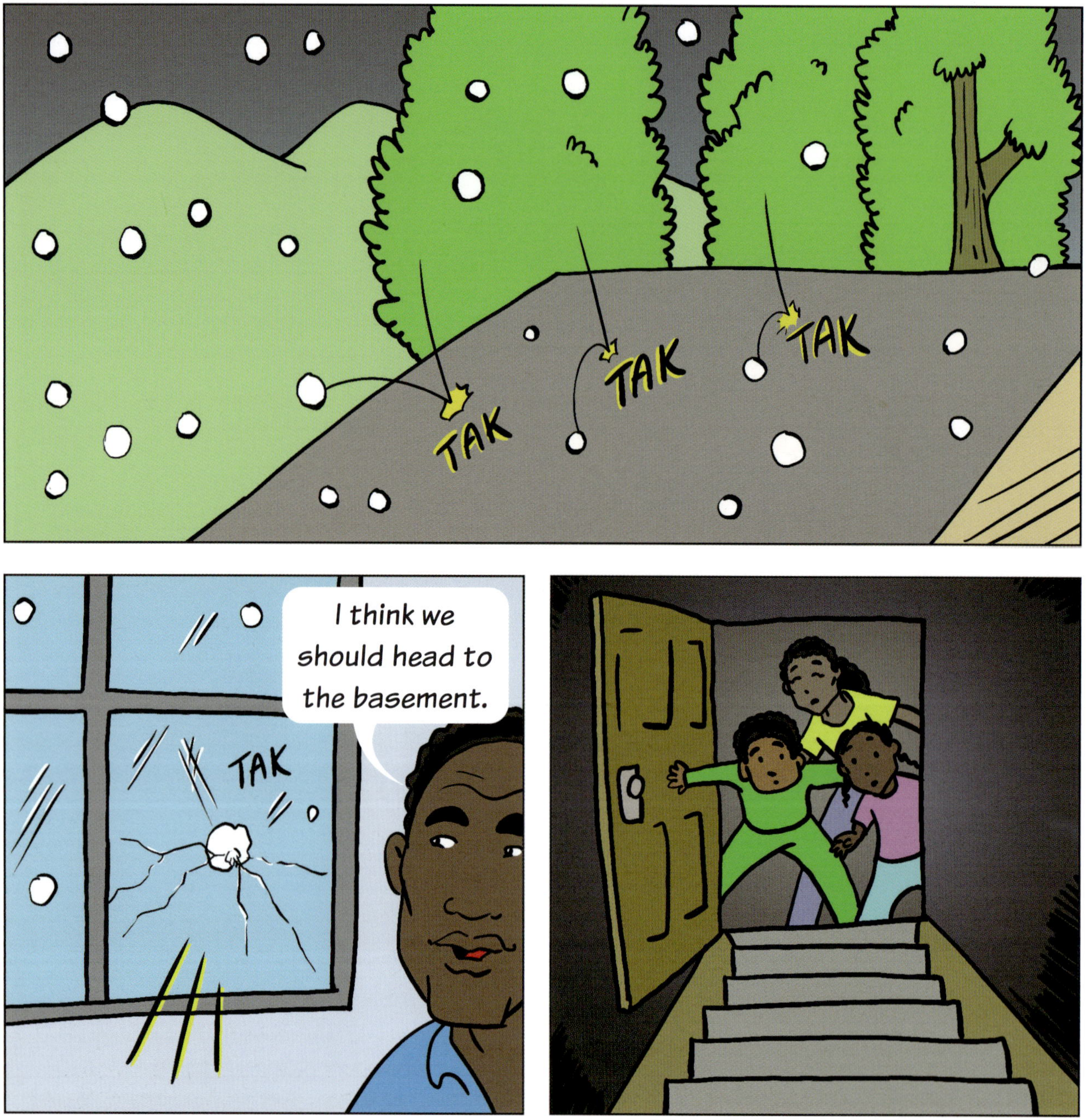
TAK
TAK
TAK
TAK
I think we should head to the basement.

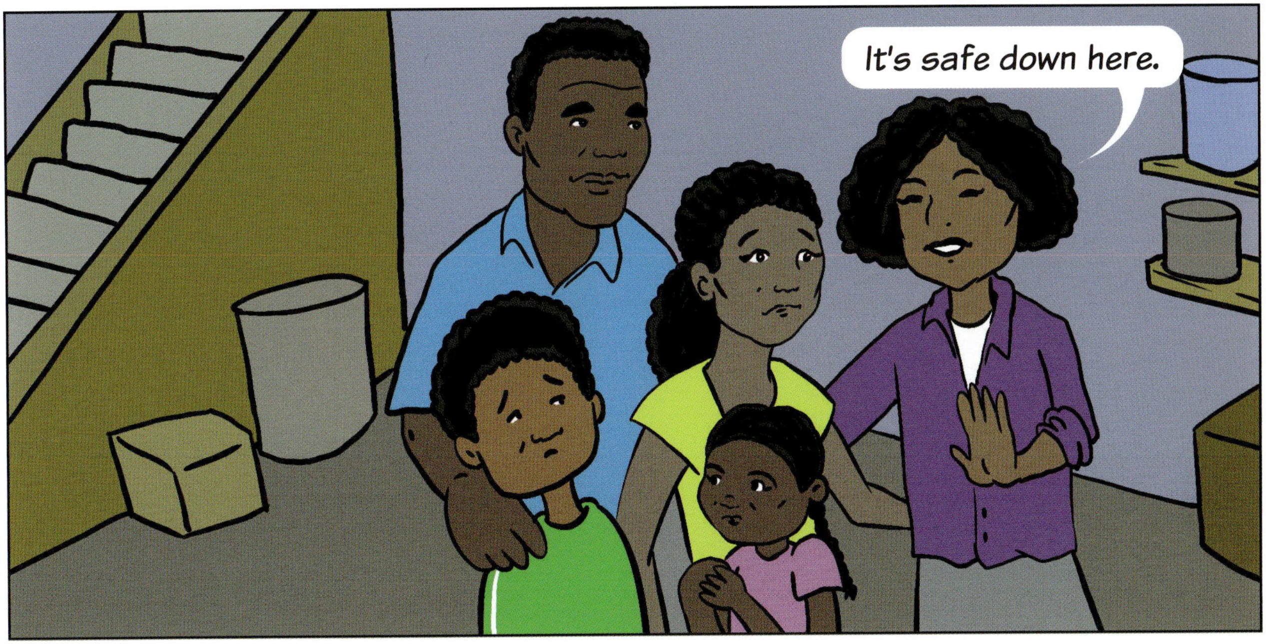
It's safe down here.

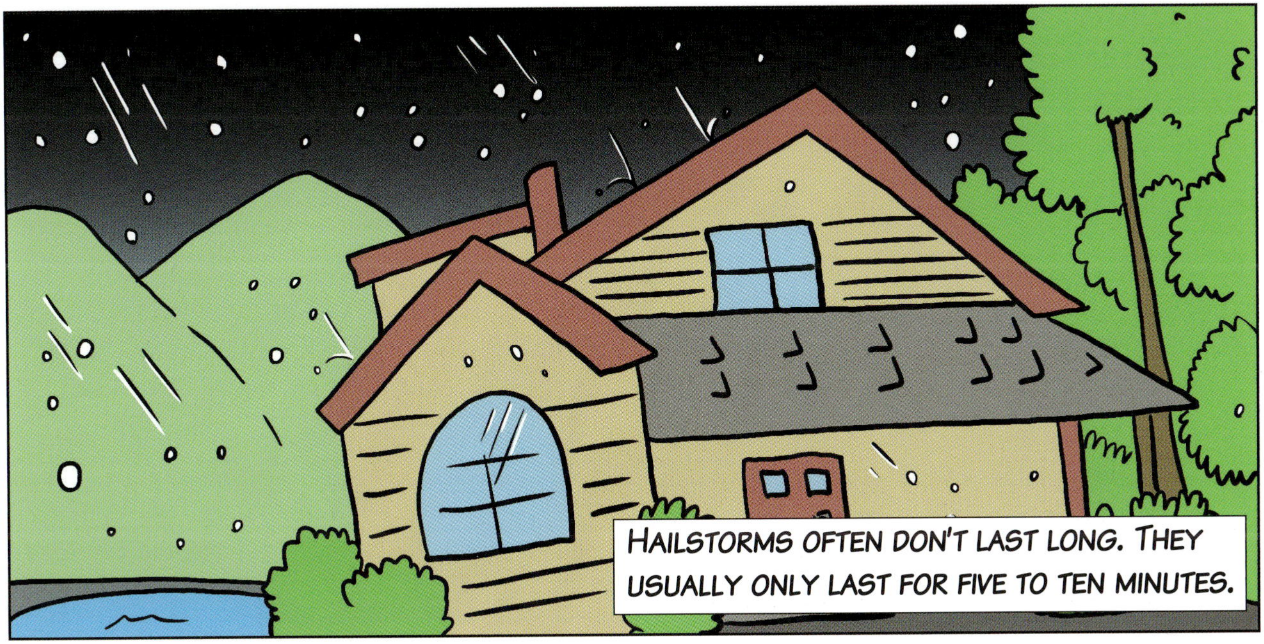
HAILSTORMS OFTEN DON'T LAST LONG. THEY USUALLY ONLY LAST FOR FIVE TO TEN MINUTES.

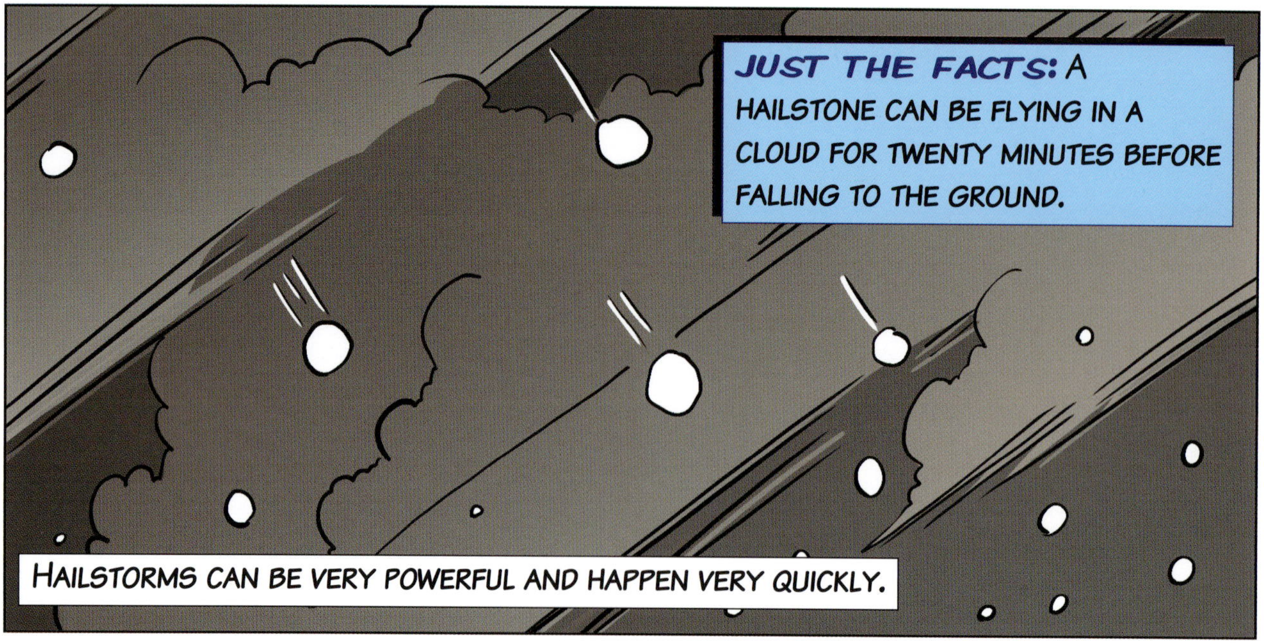
JUST THE FACTS: A HAILSTONE CAN BE FLYING IN A CLOUD FOR TWENTY MINUTES BEFORE FALLING TO THE GROUND.
HAILSTORMS CAN BE VERY POWERFUL AND HAPPEN VERY QUICKLY.

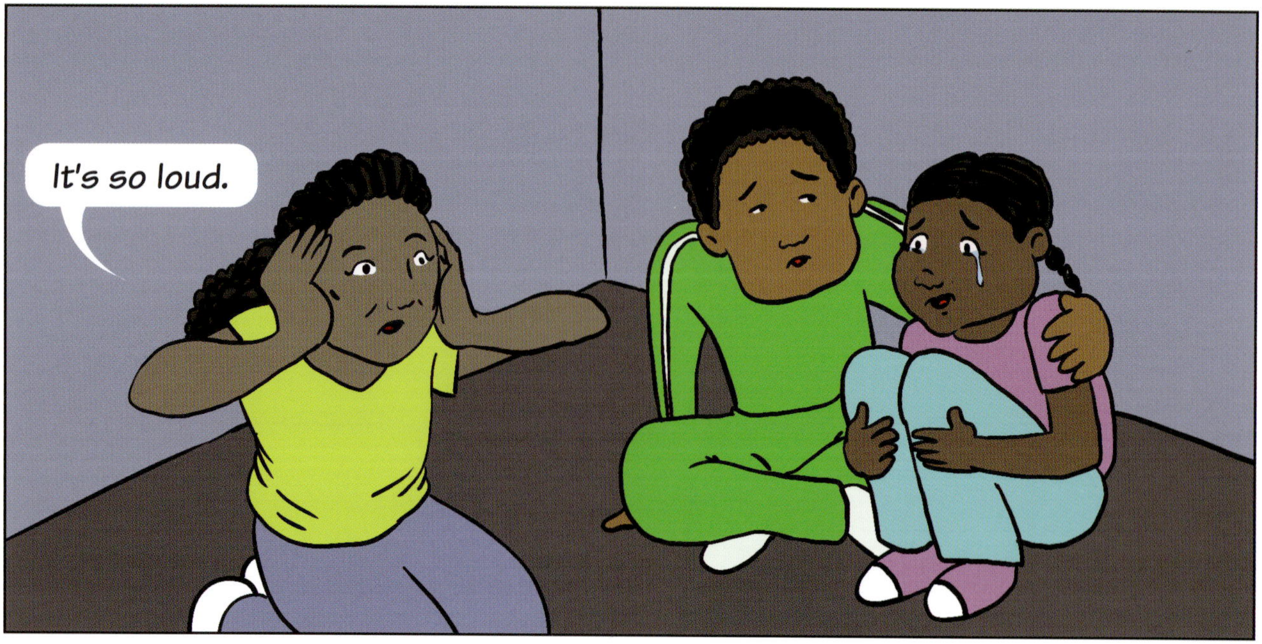
It's so loud.

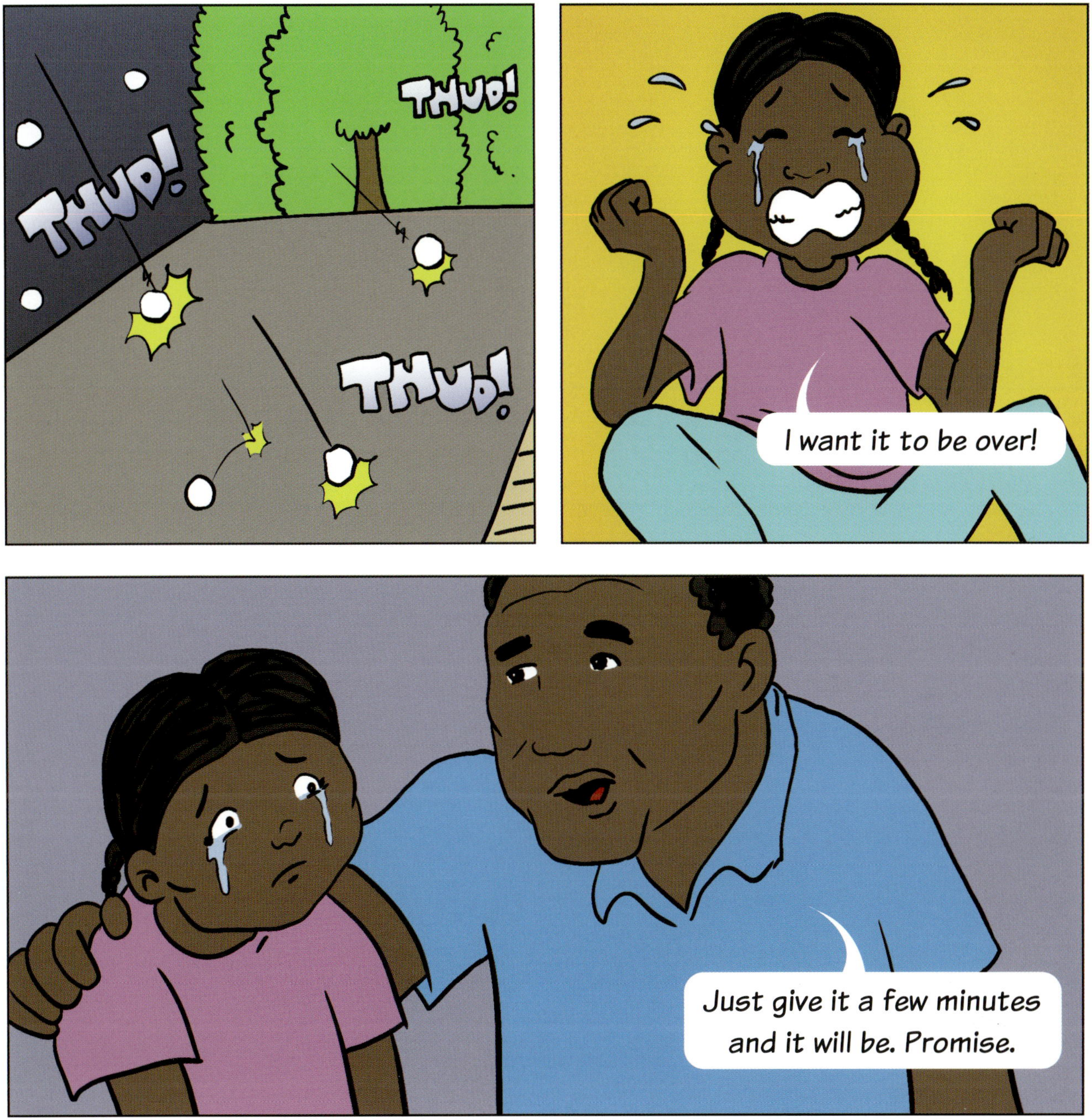
THUD!
THUD!
THUD!
I want it to be over!
Just give it a few minutes
and it will be. Promise.

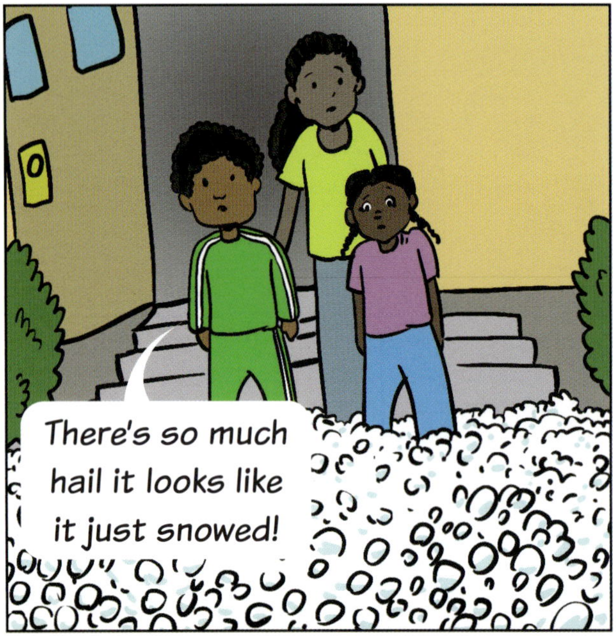
There's so much hail it looks like it just snowed!

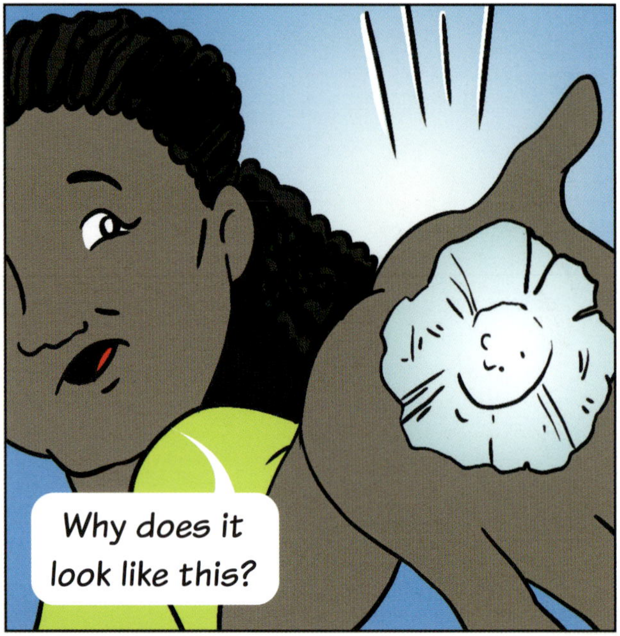
Why does it look like this?

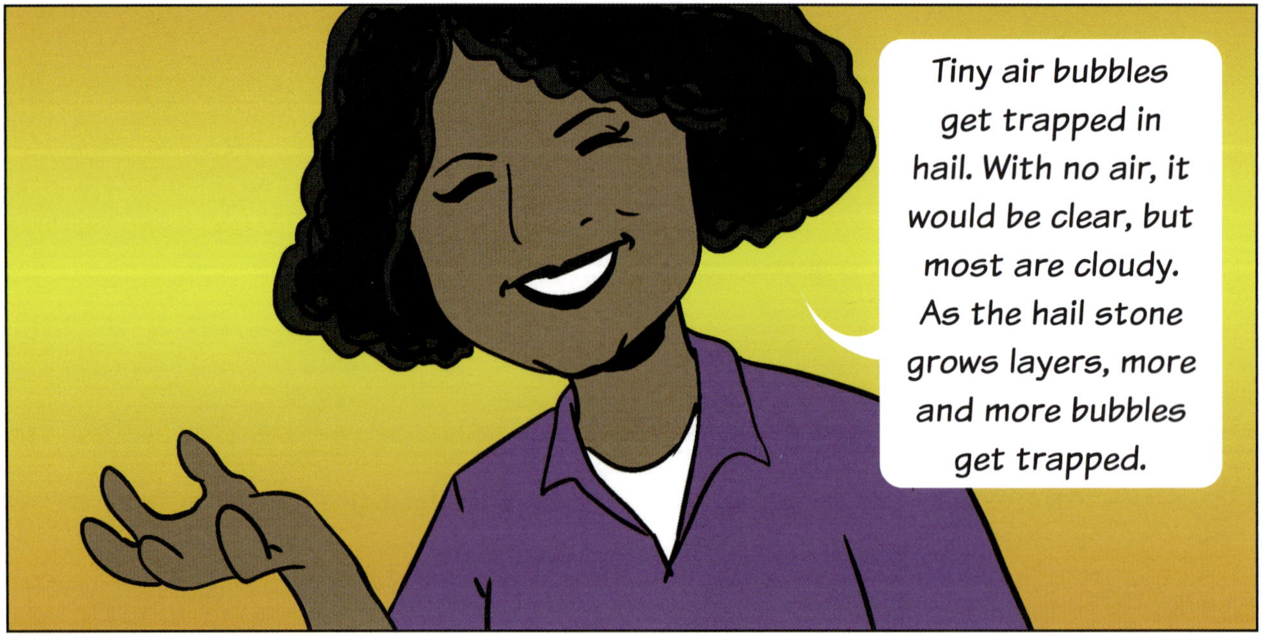
Tiny air bubbles get trapped in hail. With no air, it would be clear, but most are cloudy. As the hail stone grows layers, more and more bubbles get trapped.

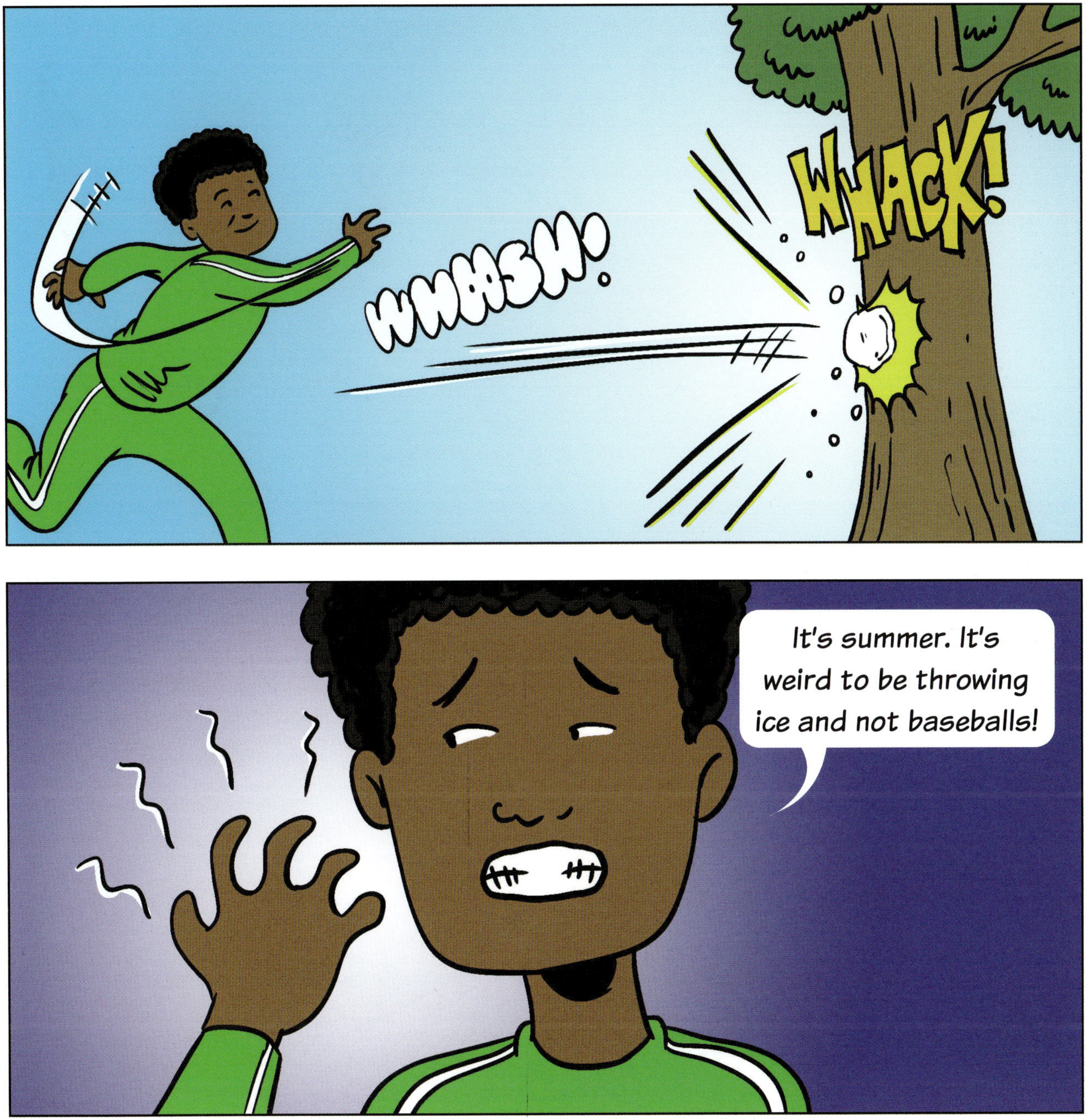
WHOOSH!
WHACK!
It's summer. It's weird to be throwing ice and not baseballs!

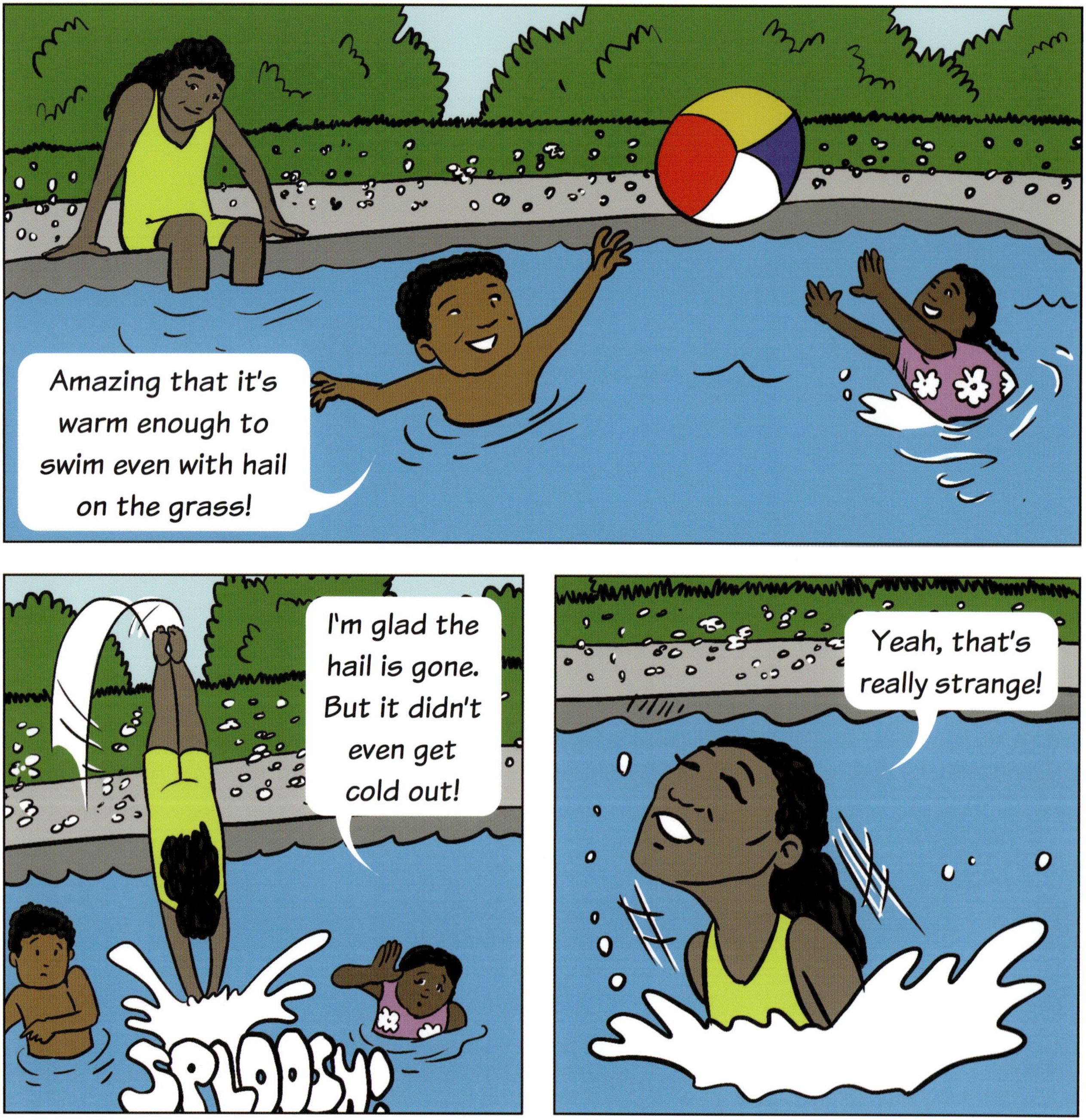
Amazing that it's warm enough to swim even with hail on the grass!
I'm glad the hail is gone. But it didn't even get cold out!
SPLOOSH!
Yeah, that's really strange!

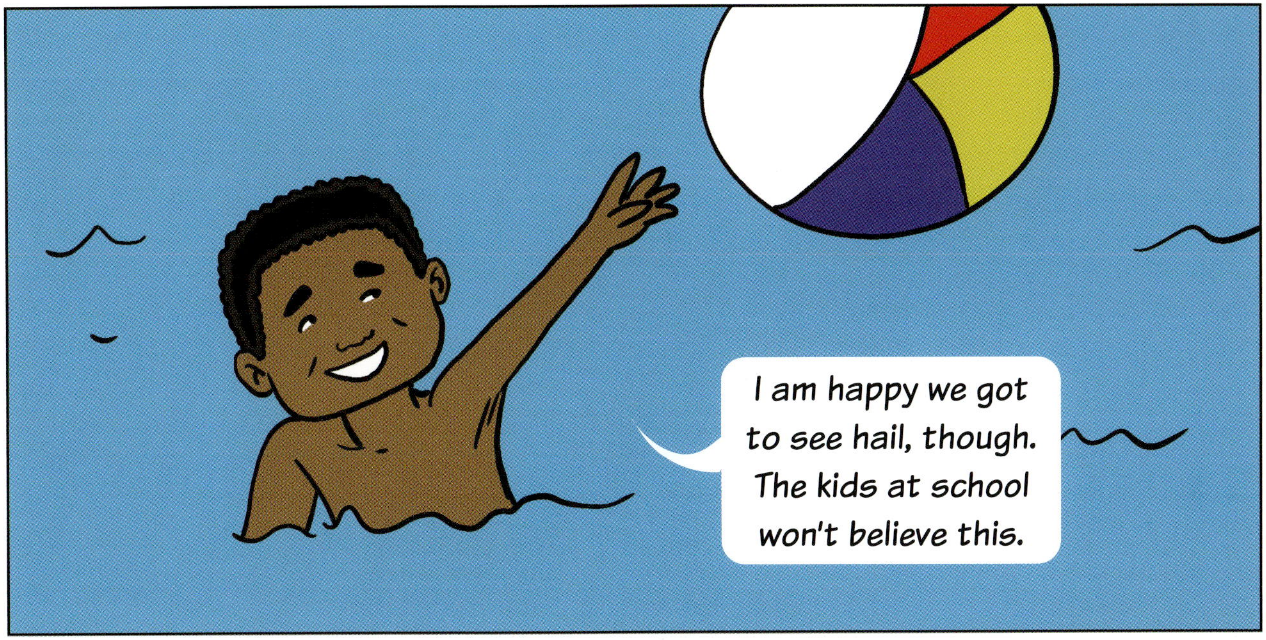
I am happy we got to see hail, though. The kids at school won't believe this.

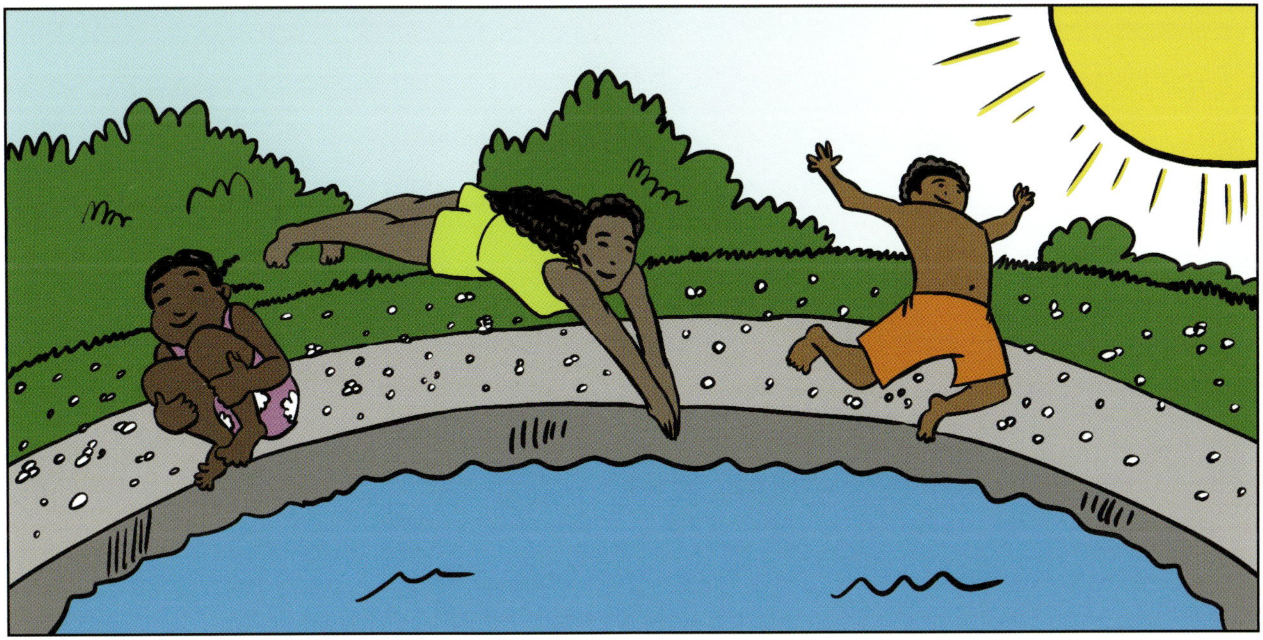

Hail Alley

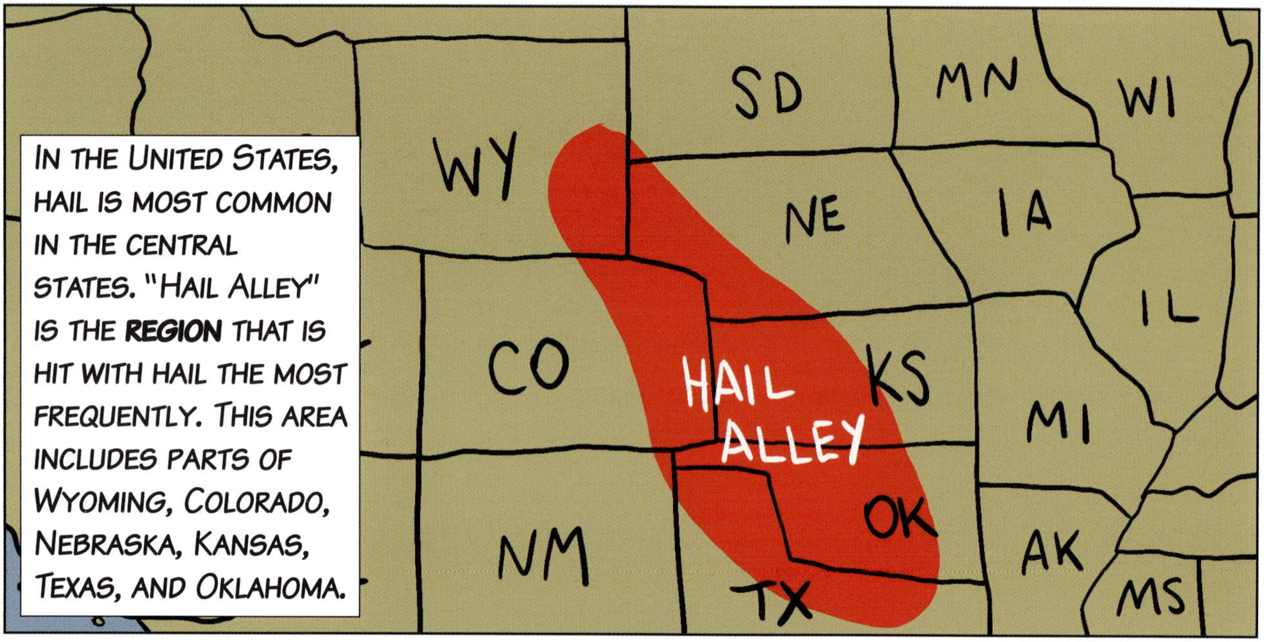
In the United States, hail is most common in the central states. "Hail Alley" is the **region** that is hit with hail the most frequently. This area includes parts of Wyoming, Colorado, Nebraska, Kansas, Texas, and Oklahoma.
SD
MN
WI
WY
NE
IA
IL
CO
HAIL ALLEY
KS
MI
NM
OK
AK
TX
MS

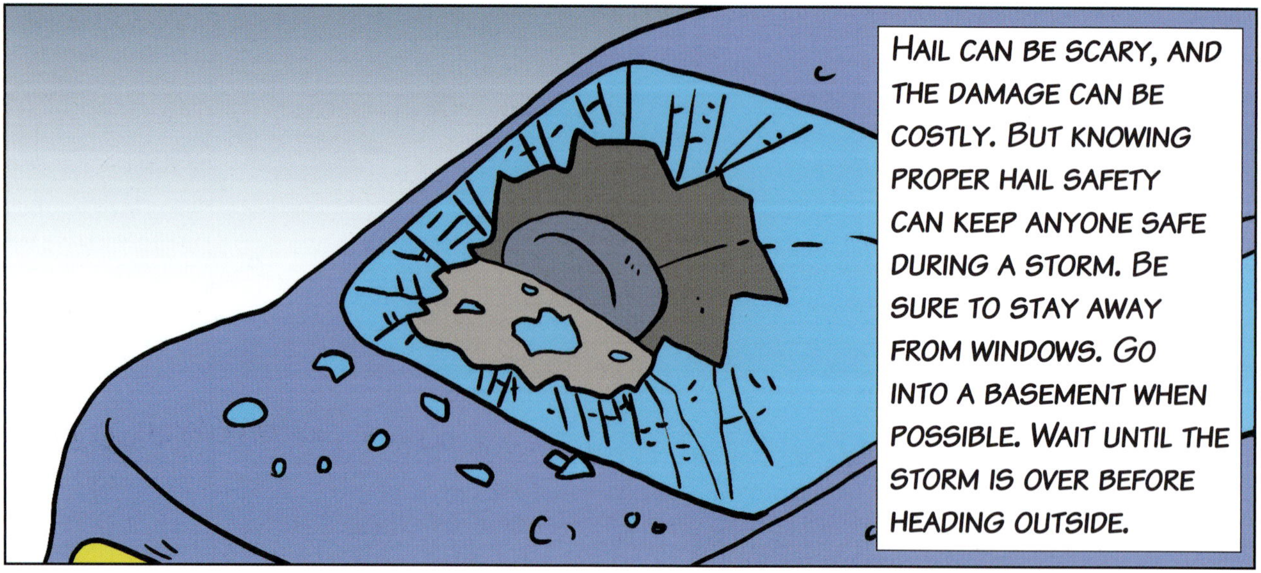
Hail can be scary, and the damage can be costly. But knowing proper hail safety can keep anyone safe during a storm. Be sure to stay away from windows. Go into a basement when possible. Wait until the storm is over before heading outside.

GLOSSARY

cattle: *cows raised for dairy products or beef*

evaporates: *changes from a liquid to a gas*

horizon: *the line where the sky and the Earth seem to meet*

particle: *a tiny piece of something*

precipitation: *water that falls from the clouds in the form of rain, hail, or snow*

radar: *a weather tool that sends out radio waves to determine the size, strength, and movement of storms*

region: *a large area*

updraft: *an upward movement of air*

FURTHER READING

Kostigen, Thomas M. *Extreme Weather: Surviving Tornadoes, Sandstorms, Hailstorms, Blizzards, Hurricanes, and More!* Washington, DC: National Geographic, 2014. This colorful book is filled with first-hand accounts of extreme storm survivors.

Probst, Jeff. *Extreme Weather.* New York: Puffin Books, 2017. This book highlights several types of extreme weather conditions and includes fun facts, quizzes, and real-life stories about extreme weather experiences.

Roker, Al. *Al Roker's Extreme Weather: Tornadoes, Typhoons, and Other Weather Phenomena.* New York: Harper Collins, 2017. This book by popular weatherman Al Roker includes a chapter about hailstorms, as well as a variety of other unique, extreme weather situations.

Hail (https://www.nationalgeographic.org/encyclopedia/hail/) This National Geographic article includes information about hailstorms, ways humans have tried to prevent hail, and a collection of photographs of hailstones.

Severe Weather 101—Hail (https://www.nssl.noaa.gov/education/svrwx101/hail/) This website published by the National Severe Storms Laboratory includes information about how hailstorms form, which areas are most likely to experience hail, and interesting facts about hailstones.

About the Author

Alexander Lowe is a writer who splits his time between Los Angeles and Chicago. He has written children's books about sports, technology, science, and media. He has also done extensive work as a sportswriter and film critic. He loves reading books of any and all kinds.

About the Illustrator

Sebastian Kadlecik is a screenwriter, actor, and comic book maker. He is best known as the creator of the epic action saga *Penguins vs. Possums*, about a secret, interspecies war for dominion over the earth, and the Eisner-nominated *Quince*, about a young Latina who gets superpowers at her quinceañera.